AF597916

Methods in Molecular Biology

For further volumes:
http://www.springer.com/series/7651

Protein Structure Prediction

Third Edition

Edited by

Daisuke Kihara

Department of Biological Sciences, Purdue University, West Lafayette, IN, USA;
Department of Computer Science, Purdue University, West Lafayette, IN, USA

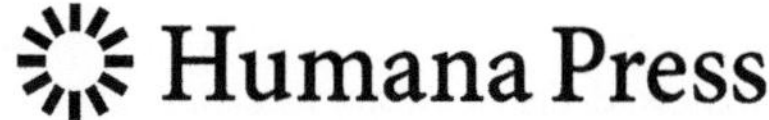

Editor
Daisuke Kihara
Department of Biological Sciences
Purdue University
West Lafayette, IN, USA

Department of Computer Science
Purdue University
West Lafayette, IN, USA

ISSN 1064-3745 ISSN 1940-6029 (electronic)
ISBN 978-1-4939-0365-8 ISBN 978-1-4939-0366-5 (eBook)
DOI 10.1007/978-1-4939-0366-5
Springer New York Heidelberg Dordrecht London

Library of Congress Control Number: 2014931090

Printed on acid-free paper

Humana Press is a brand of Springer
Springer is part of Springer Science+Business Media (www.springer.com)

Preface

Computational protein tertiary structure prediction has been extensively studied over the last two decades. Many approaches have been proposed, tested, and compared by researchers of various backgrounds who are attracted by this simple but fascinating protein structure prediction problem. We now see the field maturing, resulting in prediction methods able to produce sufficiently accurate structure models for many cases, although it is still far from the complete solution of the problem. As prediction methods become more practically useful, it becomes important to disseminate developed software to the research community so that researchers who are not necessarily familiar with computational tools can easily access and use them in their daily research activities. This software dissemination is also very beneficial for researchers who develop prediction methods because they can easily compare their methods with existing ones or integrate them as part of a pipeline of a prediction procedure. With this philosophy in mind, in the third edition of this book we focus on introducing software or Web servers which are available for researchers. This is a major difference from the second edition, which emphasized descriptive explanation of methods. Each chapter of the third edition provides practical step-by-step instructions of how to use a computational method with actual examples of prediction by the method. The chapters describe well-established methods developed by well-known researchers in this field.

The book starts by introducing three protein structure prediction/modeling methods. These methods take a single protein sequence as input and predict the tertiary structure of the input protein. The first chapter was written by Benjamin Webb and Andrej Sali on their extremely popular structure modeling tool, MODELLER. In the second chapter, Jinbo Xu and his colleagues present RaptorX, a template-based protein structure prediction server. In the third chapter, Jianlin Cheng and his colleagues provide a tutorial for their server, MULTICOM.

The next four chapters provide tools that are useful for subsequent steps of main-chain conformation prediction, which can be done by the methods introduced in Chapters 1–3. Chapter 4 deals with prediction of side-chain conformation of a protein structure model. Jiang Taijiao and his group members describe their method, RASP. The method takes the main-chain atom positions as input and builds side-chains of the protein. Chapter 5 details the use of Direct Coupling Analysis, a residue–residue contact prediction method written by Faruck Morcos, Terence Hwa, José N. Onuchic, and Martin Weigt. The method predicts physically contacting amino acid residues in the protein tertiary structure from a multiple sequence alignment. Contact prediction is useful for guiding protein structure modeling and also for selecting the most probable models from a pool of different structure models. ITScorePro, introduced in Chapter 6, is a scoring program for ranking different structure models of a target protein developed by Xiaoqin Zou's group. It is typical that a structure prediction method produces a large number of structure models, and thus identifying the most plausible model is a practically very important task in prediction. In Chapter 7, Liam James McGuffin and his colleagues describe how to use their model quality assessment server, ModFOLD. The server provides the estimated accuracy of a structure model, i.e., overall accuracy and per-residue accuracy of a model. Estimated accuracy is helpful for practical use of a model as well as for choosing the most plausible models from a pool of different models.

When the structure of a protein is modeled, it is often of interest to find similar existing structures in a database, since structure similarity often provides insight into the function and evolution of the protein. Chapter 8 introduces 3D-SURFER, a server for real-time structure database search, which was developed by Daisuke Kihara and his group members. Yaoqi Zhou and his colleagues have extended their protein structure prediction methods for predicting protein–RNA complex structures in SPOT-Seq-RNA Their method, described in Chapter 9, takes a protein sequence as input and predicts RNA sequences that would interact with the protein and the tertiary structure of the protein–RNA complex.

The subsequent two chapters are for prediction methods of intrinsic disordered regions. It has been found that many proteins have regions that are designed not to form a fixed structure and are thus called intrinsic disordered regions. These regions often have important functions such as serving as interaction sites to other proteins. In Chapter 10, Kana Shimizu describes how to use POODLE, while readers are introduced to MFDp2 by Marcin J. Mizianty, Vladimir Uversky, and Lukasz Kurgan in Chapter 11.

What follows is four chapters for protein–protein docking prediction. A protein–protein docking method predicts the complex structure of two (or more) protein structures. Chapter 12 was written by Gydo C. P. van Zudert and Alexandre M. J. J. Bonvin on their docking method, HADDOCK. In Chapter 13, the SwarmDock Web server developed by the Paul A. Bates group is introduced. The following chapter is about the DOCK/PIERR server by Ron Elber and his team. In Chapter 15, the LZerD docking program, which can perform pairwise as well as multiple protein docking, is reported by Daisuke Kihara and his lab members.

Finally, in Chapter 16, protocols for protein dynamics simulations with the CABS protein model are provided by Michal Jamroz, Andrzej Kolinski, and Sebastian Kmiecik. As you see now, this book covers a series of methods for protein structure prediction and related tools, model quality assessment, disordered region prediction, protein–protein docking, and protein dynamics. The diversity of the introduced methods shows the expansion of the computational protein structure prediction field. It will be my great pleasure if this book helps biology researchers with the use of computational methods for protein structure prediction and also serves as a bridge between computational and experimental biologists.

In closing, I would like to thank all of the authors of chapters in this book. This edition is very fortunate to have the leading experts of the field as the authors. I am also thankful to the series editor, Dr. John M. Walker for his patience and guidance and Ms. Kristen Johnson in my research group for her tremendous help in editing this book.

West Lafayette, IN, USA *Daisuke Kihara*

Contents

Contributors

BADRI ADHIKARI • *Computer Science Department, C. Bond Life Science Center, Informatics Institute, University of Missouri, Columbia, MO, USA*

PAUL A. BATES • *Biomolecular Modelling Laboratory, Cancer Research UK London Research Institute, London, UK*

DEBSWAPNA BHATTACHARYA • *Computer Science Department, C. Bond Life Science Center, Informatics Institute, University of Missouri, Columbia, MO, USA*

ALEXANDRE M.J.J. BONVIN • *Faculty of Science - Chemistry, Bijvoet Center for Biomolecular Research, Utrecht University, CH Utrecht, The Netherlands*

MARIA TERESA BUENAVISTA • *School of Biological Sciences, University of Reading, Reading, UK; BioComputing Section, Medical Research Council Harwell, Harwell Oxford, Oxfordshire, UK; Diamond Light Source, Didcot, UK*

RENZHI CAO • *Computer Science Department, C. Bond Life Science Center, Informatics Institute, University of Missouri, Columbia, MO, USA*

YANG CAO • *Key Lab of Protein and Peptide Pharmaceuticals, Institute of Biophysics, Chinese Academy of Sciences, Beijing, China*

JIANLIN CHENG • *Computer Science Department, C. Bond Life Science Center, Informatics Institute, University of Missouri, Columbia, MO, USA*

XIN DENG • *Computer Science Department, C. Bond Life Science Center, Informatics Institute, University of Missouri, Columbia, MO, USA*

JESSE EICKHOLT • *Department of Computer Science, Central Michigan University, Mt. Pleasant, MI, USA*

RON ELBER • *Department of Chemistry and Biochemistry and Institute for Computational Engineering and Sciences, University of Texas at Austin, Austin, TX, USA*

JUAN ESQUIVEL-RODRIGUEZ • *Department of Computer Science, Purdue University, West Lafayette, IN, USA*

JUAN ESQUIVEL-RODRIGUEZ • *Department of Computer Science, College of Science, Purdue University, West Lafayette, IN, USA*

VIANNEY FILOS-GONZALEZ • *Department of Mathematics, College of Science, Purdue University, West Lafayette, IN, USA*

SHENG-YOU HUANG • *Department of Physics and Astronomy, Dalton Cardiovascular Research Center, Informatics Institute; University of Missouri, Columbia, MO, USA; Department of Biochemistry, Dalton Cardiovascular Research Center, Informatics Institute; University of Missouri, Columbia, MO, USA*

TERENCE HWA • *Center for Theoretical Biological Physics, University of California at San Diego, La Jolla, CA, USA*

MICHAL JAMROZ • *Laboratory of Theory of Biopolymers, Faculty of Chemistry, University of Warsaw, Warsaw, Poland*

TAIJIAO JIANG • *Key Lab of Protein and Peptide Pharmaceuticals, Institute of Biophysics, Chinese Academy of Sciences, Beijing, China*

MORTEN KÄLLBERG • *Toyota Technological Institute, Chicago, IL, USA*

DAISUKE KIHARA • *Department of Biological Sciences, Purdue University, West Lafayette, IN, USA; Department of Computer Science, Purdue University, West Lafayette, IN, USA*

SEBASTIAN KMIECIK • *Laboratory of Theory of Biopolymers, Faculty of Chemistry, University of Warsaw, Warsaw, Poland*
ANDRZEJ KOLINSKI • *Laboratory of Theory of Biopolymers, Faculty of Chemistry, University of Warsaw, Warsaw, Poland*
LUKASZ KURGAN • *Department of Electrical and Computer Engineering, University of Alberta, Edmonton, Canada*
JILONG LI • *Computer Science Department, C. Bond Life Science Center, Informatics Institute, University of Missouri, Columbia, MO, USA*
BIN LI • *La Jolla Institute for Allergy and Immunology, San Diego, CA, USA*
JIANZHU MA • *Toyota Technological Institute, Chicago, IL, USA*
GOHAR MARGARYAN • *Toyota Technological Institute, Chicago, IL, USA*
LIAM JAMES MCGUFFIN • *School of Biological Sciences, University of Reading, Reading, UK*
ZHICHAO MIAO • *Key Lab of Protein and Peptide Pharmaceuticals, Institute of Biophysics, Chinese Academy of Sciences, Beijing, China*
MARCIN J. MIZIANTY • *Department of Electrical and Computer Engineering, University of Alberta, Edmonton, Canada*
FARUCK MORCOS • *Center for Theoretical Biological Physics, Rice University, Houston, TX, USA*
JOSÉ N. ONUCHIC • *Center for Theoretical Biological Physics, Rice University, Houston, TX, USA*
D.V.S. RAVIKANT • *WalmartLabs, San Bruno, CA, USA*
DANIEL BARRY ROCHE • *Genoscope, Institut de Génomique, Commissariat à l'Energie Atomique et aux Energies Alternatives, Evry, France; Centre National de la Recherche Scientifique, UMR Evry, Evry, France; Université d'Evry-Val-d'Essonne, Evry, France; PRES UniverSud Paris, Les Algorithmes, Bâtiment Euripide, Saint-Aubin, France*
LEE SAEL • *Department of Computer Science, State University of New York Korea, Incheon, South Korea*
ANDREJ SALI • *Department of Bioengineering and Therapeutic Sciences, University of California, San Francisco, CA, USA; Department of Pharmaceutical Chemistry, California Institute for Quantitative Biosciences (QB3), University of California San Francisco, San Francisco, CA, USA*
KANA SHIMIZU • *Computational Biology Research Center, National Institute of Advanced Industrial Science and Technology, Tokyo, Japan*
MIECZYSLAW TORCHALA • *Biomolecular Modelling Laboratory, Cancer Research UK London Research Institute, London, UK*
VLADIMIR UVERSKY • *Department of Molecular Medicine, Byrd Alzheimer's Research Institute, Morsani College of Medicine, University of South Florida, Tampa, FL, USA; Institute for Biological Instrumentation, Russian Academy of Sciences, Moscow Region, Russia*
SHRUTHI VISWANATH • *Department of Computer Science, University of Texas at Austin, Austin, TX, USA; Institute for Computational Engineering and Sciences, University of Texas at Austin, Austin, TX, USA*
SHENG WANG • *Toyota Technological Institute, Chicago, IL, USA*
JIHUA WANG • *Shandong Provincial Key Laboratory of Functional Macromolecular Biophysics and Department of Physics, Dezhou University, Dezhou, China*
BENJAMIN WEBB • *Department of Bioengineering and Therapeutic Sciences, University of California, San Francisco, CA, USA; Department of Pharmaceutical Chemistry,*

California Institute for Quantitative Biosciences (QB3), University of California San Francisco, San Francisco, CA, USA

MARTIN WEIGT • *UMR7238—Laboratoire de Génomique des Microorganismes, Université Pierre et Marie Curie, Paris, France*

YI XIONG • *Department of Biological Sciences, Purdue University, West Lafayette, IN, USA*

JINBO XU • *Toyota Technological Institute, Chicago, IL, USA*

YUEDONG YANG • *School of Informatics, Indiana University Purdue University, Indianapolis, IN, USA; Center for Computational Biology and Bioinformatics, Indiana University School of Medicine, Indianapolis, IN, USA; Institute for Glycomics and School of Information and Communication Technology, Griffith University, Southport, QLD, Australia*

HUIYING ZHAO • *School of Informatics, Indiana University Purdue University, Indianapolis, IN, USA; Center for Computational Biology and Bioinformatics, Indiana University School of Medicine, Indianapolis, IN, USA*

YAOQI ZHOU • *School of Informatics, Indiana University Purdue University, Indianapolis, IN, USA; Center for Computational Biology and Bioinformatics, Indiana University School of Medicine, Indianapolis, IN, USA; Institute for Glycomics and School of Information and Communication Technology, Griffith University, Southport, QLD, Australia*

XIAOQIN ZOU • *Department of Physics and Astronomy, Dalton Cardiovascular Research Center, Informatics Institute; University of Missouri, Columbia, MO, USA; Department of Biochemistry, Dalton Cardiovascular Research Center, Informatics Institute; University of Missouri, Columbia, MO, USA*

GYDO C.P. VAN ZUNDERT • *Faculty of Science - Chemistry, Bijvoet Center for Biomolecular Research, Utrecht University, CH Utrecht, The Netherlands*

Chapter 1

Protein Structure Modeling with *MODELLER*

Benjamin Webb and Andrej Sali

Abstract

Genome sequencing projects have resulted in a rapid increase in the number of known protein sequences. In contrast, only about one-hundredth of these sequences have been characterized at atomic resolution using experimental structure determination methods. Computational protein structure modeling techniques have the potential to bridge this sequence–structure gap. In this chapter, we present an example that illustrates the use of MODELLER to construct a comparative model for a protein with unknown structure. Automation of a similar protocol has resulted in models of useful accuracy for domains in more than half of all known protein sequences.

Key words Comparative modeling, Fold assignment, Sequence–structure alignment, Model assessment, Multiple templates

1 Introduction

The function of a protein is determined by its sequence and its three-dimensional (3D) structure. Large-scale genome sequencing projects are providing researchers with millions of protein sequences, from various organisms, at an unprecedented pace. However, the rate of experimental structural characterization of these sequences is limited by the cost, time, and experimental challenges inherent in the structural determination by X-ray crystallography and nuclear magnetic resonance (NMR) spectroscopy.

In the absence of experimentally determined structures, computationally derived protein structure models are often valuable for generating testable hypotheses [1, 2]. Such models are generally produced using either comparative modeling methods or free modeling techniques (also referred to as ab initio or de novo modeling) [3]. Comparative modeling relies on structural information from related proteins to guide the modeling procedure [4–6]. Free modeling does not require a related protein but instead uses a variety of methods to combine physics with the known behaviors of protein structures (for example by combining multiple short

Daisuke Kihara (ed.), *Protein Structure Prediction*, Methods in Molecular Biology, vol. 1137,
DOI 10.1007/978-1-4939-0366-5_1,

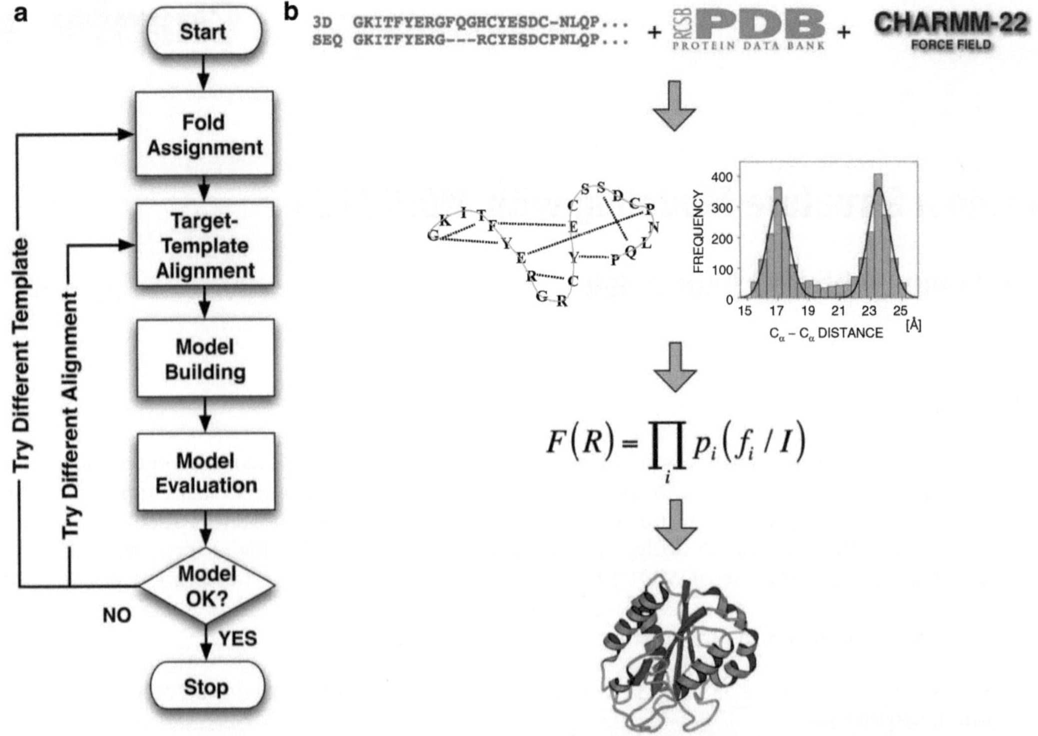

Fig. 1 Comparative protein structure modeling. (**a**) A flow chart illustrating the steps in the construction of a comparative model [4]. (**b**) Description of comparative modeling by extraction of spatial restraints as implemented in MODELLER [12]. By default, spatial restraints in MODELLER involve (1) homology-derived restraints from the aligned template structures, (2) statistical restraints derived from all known protein structures, and (3) stereochemical restraints from the CHARMM-22 molecular mechanics force field. These restraints are combined into an objective function that is then optimized to calculate the final 3D model of the target sequence

structural fragments extracted from known proteins) [7–9]; it is, however, extremely computationally expensive [3]. Comparative protein structure modeling, which this text focuses on, has been used to produce reliable structure models for at least one domain in more than half of all known sequences [10]. Hence, computational approaches can provide structural information for two orders of magnitude more sequences than experimental methods and are expected to be increasingly relied upon as the gap between the number of known sequences and the number of experimentally determined structures continues to widen.

Comparative modeling consists of four main steps [4] (Fig. 1): (1) fold assignment that identifies overall similarity between the target sequence and at least one known structure (template); (2) alignment of the target sequence and the template(s); (3) building a model based on the alignment with the chosen template(s); and (4) predicting the accuracy of the model.

MODELLER is a computer program for comparative protein structure modeling [11, 12]. In the simplest case, the input is an alignment of a sequence to be modeled with the template structure(s), the atomic coordinates of the template(s), and a simple script file. MODELLER then automatically calculates a model containing all non-hydrogen atoms, without any user intervention and within minutes on a desktop computer. Apart from model building, MODELLER can perform auxiliary tasks such as fold assignment, alignment of two protein sequences or their profiles [13], multiple alignment of protein sequences and/or structures [14, 15], clustering of sequences and/or structures, and ab initio modeling of loops in protein structures [11].

MODELLER implements comparative protein structure modeling by satisfaction of spatial restraints that include (1) homology-derived restraints on the distances and dihedral angles in the target sequence, extracted from its alignment with the template structures [12]; (2) stereochemical restraints such as bond length and bond angle preferences, obtained from the CHARMM-22 molecular mechanics force field [16]; (3) statistical preferences for dihedral angles and non-bonded interatomic distances, obtained from a representative set of known protein structures [17, 18]; and (4) optional manually curated restraints, such as those from NMR spectroscopy, rules of secondary structure packing, cross-linking experiments, fluorescence spectroscopy, image reconstruction from electron microscopy, site-directed mutagenesis, and intuition (Fig. 1). The spatial restraints, expressed as probability density functions, are combined into an objective function that is optimized by a combination of conjugate gradients and molecular dynamics with simulated annealing. This model building procedure is similar to structure determination by NMR spectroscopy.

In this chapter, we use a sequence with unknown structure to illustrate the use of various modules in MODELLER to perform the four steps of comparative modeling.

2 Materials

To follow the examples in this discussion, both the MODELLER software and a set of suitable input files are needed. The MODELLER software is free for academic use; it can be downloaded from http://salilab.org/modeller/ and is available in binary form for most common machine types and operating systems (*see* **Note 1**). This text uses MODELLER 9.11, the most recent version at the time of writing, but the examples should also work with any newer version. The example input files can be downloaded from http://salilab.org/modeller/tutorial/MMB13.zip.

All MODELLER scripts are Python scripts. Python is pre-installed on most Linux and Mac machines; Windows users can obtain it from http://www.python.org/. It is not necessary to install Python, or to have a detailed knowledge of its use, to use MODELLER, but it is helpful for creating and understanding the more advanced MODELLER scripts.

3 Methods

The procedure for calculating a 3D model for a sequence with unknown structure will be illustrated using the following example: a novel gene for lactate dehydrogenase (LDH) was identified from the genomic sequence of *Trichomonas vaginalis* (TvLDH). The corresponding protein had higher sequence similarity to the malate dehydrogenase of the same species (TvMDH) than to any other LDH [19]. Comparative models were constructed for TvLDH and TvMDH to study the sequences in a structural context and to suggest site-directed mutagenesis experiments to elucidate changes in enzymatic specificity in this apparent case of convergent evolution. The native and mutated enzymes were subsequently expressed and their activities compared [19].

`Monospaced text` is used below for computer file and folder/directory names, command lines, file contents, and variable and class names.

3.1 Fold Assignment

The first step in comparative modeling is to identify one or more templates (sequences with known 3D structure) for the modeling procedure. One way to do this is to search a database of experimentally determined structures extracted from the Protein Data Bank (PDB) [20] to find sequences that have detectable similarity to the target (*see* **Note 2**). To prepare this database (*see* **Note 3**), run the following command from the command line (*see* **Note 4**):

```
python make_pdb_95.py>make_pdb_95.log
```

This generates a file called `pdb_95.bin`, which is a binary representation of the search database (*see* **Note 5**), and a log file, `make_pdb_95.log`. Next, MODELLER's `profile.build()` command is used; this uses the local dynamic programming algorithm to identify sequences related to TvLDH [21]. In the simplest case, `profile.build()` takes as input the target sequence, in file `TvLDH.ali` (*see* **Note 6**), and the binary database and returns a set of statistically significant alignments (file `build_profile.prf`) and a MODELLER log file (`build_profile.log`). Run this step by typing

```
python build_profile.py>build_profile.log
```

The first few lines of the resulting `build_profile.prf` will look similar to (*see* **Note 7**) the following (note that the rightmost column, containing the primary sequence, has been omitted here for clarity):

```
# Number of sequences:     51
# Length of profile  :    335
# N_PROF_ITERATIONS  :      1
# GAP_PENALTIES_1D   :   -500.0   -50.0
# MATRIX_OFFSET      : -450.0
# RR_FILE            : ${LIB}/blosum62.sim.mat
 1 TvLDH   S  0  335   1   335   0    0    0  0.  0.0
 2 1a5zA   X  1  312  75   242  63  229  164 28.  0.34E-07
 3 2a92A   X  1  316   8   191   6  186  174 26.  0.69E-04
 4 4aj2A   X  1  327  85   301  89  300  207 25.  0.15E-04
 5 1b8pA   X  1  327   7   331   6  325  316 42.  0.0
```

The first six lines of this file contain the input parameters used to create the alignments. Subsequent lines contain several columns of data; for the purposes of this example, the most important columns are (1) the second column, containing the PDB code of the related template sequences; (2) the eleventh column, containing the percentage sequence identity between the TvLDH and template sequences; and (3) the twelfth column, containing the *E*-values for the statistical significance of the alignments. These columns are shown in bold above.

The extent of similarity between the target–template pairs is usually quantified using sequence identity or a statistical measure such as *E*-value (*see* **Note 8**). Inspection of column 11 shows that the template with the highest sequence identity with the target is the 1y7tA structure (45 % sequence identity). Further inspection of column 12 shows that there are nine PDB sequences, all corresponding to malate dehydrogenases (1b8pA, 1civA, 3d5tA, 4h7pA, 4h7pB, 5mdhA, 7mdhA, 1smkA, 1y7tA) that show significant similarities to TvLDH with *E*-values of zero.

3.2 Sequence–Structure Alignment

The next step is to align the target TvLDH sequence with the chosen template (*see* **Note 9**). Here, the 1y7tA template is used. This alignment is created using MODELLER's `align2d()` function (*see* **Note 10**). Although `align2d()` is based on a global dynamic programming algorithm [22], it is different from standard sequence–sequence alignment methods because it takes into account structural information from the template when constructing an alignment. This task is achieved through a variable gap penalty function that tends to place gaps in solvent-exposed and curved regions, outside secondary structure segments, and not between two positions that are close in space [14]. In the current example, the target–template similarity is so high that almost any method with reasonable parameters will result in the correct alignment (*see* **Note 11**).

This step is carried out by running

```
python align2d.py>align2d.log
```

This script reads in the PDB structure of the template and the sequence of the target (TvLDH) and calls the `align2d()` function to perform the alignment. The resulting alignment is written out in two formats. `TvLDH-1y7tA.ali` in the PIR format is subsequently used by MODELLER for modeling; `TvLDH-1y7tA.pap` in the PAP format is easier to read, for example to see which residues are aligned with each other.

3.3 Model Building

Models of TvLDH can now be built by running

```
python model.py>model.log
```

The script uses MODELLER's `automodel` class, specifying the name of the alignment file to use and the identifiers of the target (TvLDH) and template (1y7tA) sequences. It then asks `automodel` to generate five models (*see* **Note 12**). Each is assessed with the normalized Discrete Optimized Protein Energy (DOPE) assessment method [18]. The five models are written out as PDB files with names `TvLDH.B9999[0001-0005].pdb`.

3.4 Model Evaluation

The log file produced by the model building procedure (`model.log`) contains a summary of each calculation at the bottom of the file. This summary includes, for each of the five models, the MODELLER objective function (*see* **Note 13**) [12] and the normalized DOPE score (*see* **Note 14**). These scores can be used to identify which of the five models produced is likely to be the most accurate model (*see* **Note 15**).

Since the DOPE potential is simply a sum of interactions between pairs of atoms, it can be decomposed into a score per residue, which is termed in MODELLER an "energy profile." This energy profile can be generated for the model with the best DOPE score by running the `make_energy_profile.py` script. The script outputs the profile, `TvLDH.profile`, in a simple format that is easily displayed in any graphing package. Such a profile is useful to detect local regions of high pseudo-energy that usually correspond to errors in the model (*see* **Notes 16** and **17**).

3.5 Use of Multiple Templates

One way to potentially improve the accuracy of generated models is to use multiple-template structures. When there are multiple templates, different template structures may be of higher local sequence identity to the target (or higher quality) than others in different regions, allowing MODELLER to build a model based on the most useful structural information for each region in the protein. The procedure is demonstrated here using the five templates that have the highest sequence identity to the target (1b8pA, 4h7pA, 4h7pB, 5mdhA, 1y7tA). Input files can be found in the "`multiple`" subdirectory of the zipfile. The first step is to align all of the templates with each other, which can be done by running

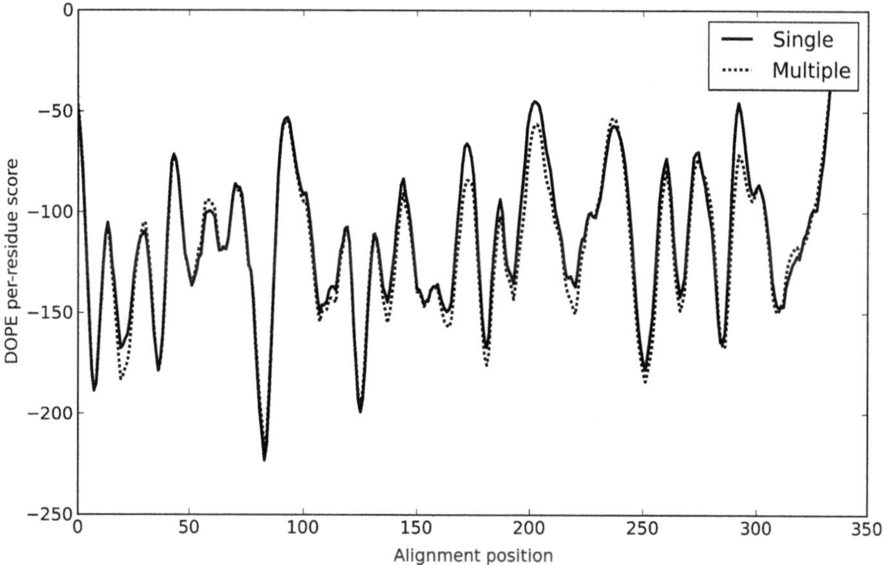

Fig. 2 The DOPE [18] energy profiles for the best-assessed model generated by modeling with a single template (*solid line*) and multiple templates (*dotted line*). Peaks (local regions of high, unfavorable score) tend to correspond to errors in the models

```
python salign.py>salign.log
```

This script uses MODELLER's `salign()` function [15] to read in all of the template structures and then generate their best structural alignment (*see* **Note 18**), written out as `templates.ali`.

Next, just as for single-template modeling, the target is aligned with the templates using the `align2d()` function. The function's `align_block` parameter is set to 5 to align the target sequence with the pre-aligned block of templates, and not to change the existing alignment between individual templates:

```
python align2d.py>align2d.log
```

Finally, model generation proceeds just as for the single-template case (the only difference is that `automodel` is now given a list of all five templates):

```
python model.py>model.log
```

Comparison of the normalized DOPE scores from the end of this logfile with those from the single-template case shows an improvement in the DOPE score of the best model from −0.92 to −1.19. Figure 2 shows the energy profiles of the best scoring models from each procedure (generated using the `plot_profiles.py` script). It can be seen that some of the predicted errors in the single-template model (peaks in the graph) have been resolved in the model calculated using multiple templates.

3.6 External Assessment

Models generated by MODELLER are stored in PDB files and so can be evaluated for accuracy with other methods if desired. One such method is the ModEval web server at http://salilab.org/evaluation/. This server takes as input the PDB file and the MODELLER PIR alignment used to generate it. It returns not only the normalized DOPE score and the energy profile but also the GA341 assessment score [23, 24] and an estimate of the Cα RMSD and native overlap between the model and its hypothetical native structure, using the TSVMod method [25]; native overlap is defined as the fraction of Cα atoms in the model that are within 3.5 Å of the same Cα atom in the native structure after least squares superposition.

3.7 Structures of Complexes

The example shown here generates a model of a single protein. However, MODELLER can also generate models of complexes of multiple proteins if templates for the entire complex are available; examples can be found in the MODELLER manual. In the case where only templates for the individual subunits in the complex are available, comparative models can be docked in a pairwise fashion by molecular docking [26, 27] or assembled based on various experimental data to generate approximate models of the complex using a wide variety of integrative modeling methods [28–31]. For example, if a cryo-electron microscopy density map of the complex is available, a model of the whole complex can be constructed by simultaneously fitting comparative models of the subunits into the density map using the MultiFit method [32] or its associated web server at http://salilab.org/multifit/ [33]. Alternatively, if a small-angle X-ray (SAXS) profile of a dimer is available, models of the dimer can be generated by docking the two subunits, constrained by the SAXS data, using the FoXSDock web server at http://salilab.org/foxsdock/ [34]. Both of these methods are part of the open-source *Integrative Modeling Platform* (IMP) package [29].

4 Notes

1. The MODELLER website also contains a full manual, a mailing list, and more example MODELLER scripts. A license key is required to use MODELLER, but this can also be obtained from the website.
2. The sequence identity is a useful predictor of the accuracy of the final model when its value is >30 %. It has been shown that models based on such alignments usually have, on average, more than ~60 % of the backbone atoms correctly modeled with a root-mean-squared deviation (RMSD) for Cα atoms of less than 3.5 Å (Fig. 3). Sequence–structure relationships in the "twilight zone" [35] (corresponding to relationships with statistically significant sequence similarity with identities

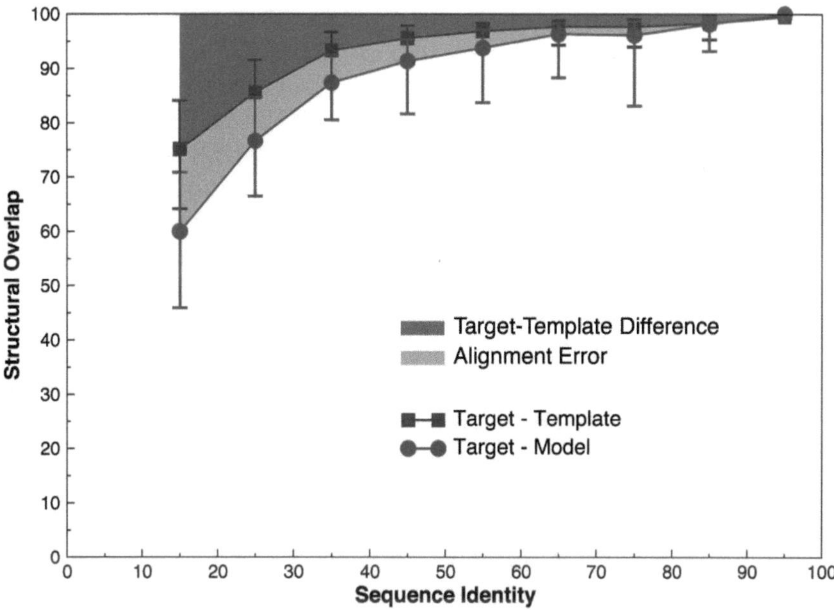

Fig. 3 Average model accuracy as a function of sequence identity [54]. As the sequence identity between the target sequence and the template structure decreases, the average structural similarity between the template and the target also decreases (*dark grey area, squares*) [55]. Structural overlap is defined as the fraction of equivalent Cα atoms. For the comparison of the model with the actual structure (*circles*), two Cα atoms were considered equivalent if they belonged to the same residue and were within 3.5 Å of each other after least squares superposition. For comparisons between the template structure and the actual target structure (*squares*), two Cα atoms were considered equivalent if they were within 3.5 Å of each other after alignment and rigid-body superposition. The difference between the model and the actual target structure is a combination of the target–template differences (*dark grey area*) and the alignment errors (*light grey area*). The figure was constructed by calculating ~1 million comparative models based on single template of varying similarity to the targets. All targets had known (experimentally determined) structures

generally in the 10–30 % range), or the "midnight zone" [35] (corresponding to statistically insignificant sequence similarity), typically result in less accurate models.

3. The database contains sequences of the structures from PDB. To increase the search speed, redundancy is removed from the database; the PDB sequences are clustered with other sequences that are at least 95 % identical, and only the representative of each cluster is stored in the database. This database is termed "pdb_95." A copy of this database is included in the downloaded zipfile as `pdb_95.pir`. Newer versions of this database, updated as new structures are deposited in PDB, can be downloaded from the MODELLER website at http://salilab.org/modeller/supplemental.html.
4. MODELLER is a command line tool, and so all commands must be run by typing at the command line. All of the necessary

input files for this demonstration are in the downloaded zipfile; simply download and extract the zipfile and change into the newly created directory (using the "cd" command at the command line). After this, MODELLER scripts can be run as shown in the text. All MODELLER scripts are Python scripts and so should be run with the "python" command. (On some systems the full path to the Python interpreter is necessary, such as /usr/bin/python on a Linux or a Mac machine or C:\python26\python.exe on a Windows system.) MODELLER scripts can also be run from other Python frontends, such as IDLE, if desired. On a Windows system, it is generally *not* a good idea to simply "double click" on a MODELLER Python script, since any output from the script will disappear as soon as it finishes. Finally, if Python is not installed, MODELLER includes a basic Python 2.3 interpreter as "mod<version>." For example, to run this first script using MODELLER version 9.11's own interpreter, run "mod9.11 make_pdb_95.py." Note that mod9.11 automatically creates a "make_pdb_95.log" logfile.

5. The binary database is much faster to use than the original text format database, pdb_95.pir. Note, however, that it is not necessarily smaller. This script does not need to be run again unless pdb_95.pir is updated.
6. TvLDH.ali simply contains the primary sequence of the target, in MODELLER's variant of the PIR format (which is documented in more detail in the MODELLER manual). This file is included in the zipfile.
7. Although MODELLER's algorithms are deterministic, exactly the same job run on different machines (e.g., a Linux box versus a Windows or a Mac machine) may give different results. This difference may arise because different machines handle rounding of floating point numbers and ordering of floating point operations differently, and the minor differences introduced can be compounded and end up giving very different outputs. This variation is normal and to be expected, and so the results shown in this text may differ from those obtained by running MODELLER elsewhere.
8. The sequence identity is not a statistically reliable measure of alignment significance and corresponding model accuracy for values lower than 30 % [35, 36]. During a scan of a large database, for instance, it is possible that low values occur purely by chance. In such cases, it is useful to quantify the sequence–structure relationship using more robust measures of statistical significance, such as *E*-values [37], that compare the score obtained for an alignment with an established background distribution of such scores. One other problem of using sequence identity as a measure to select templates is that, in practice,

there is no single generally used way to normalize it [36]. For instance, local alignment methods usually normalize the number of identically aligned residues by the length of the alignment, while global alignment methods normalize it by either the length of the target sequence or the length of the shorter of the two sequences. Therefore, it is possible that alignments of short fragments produce a high sequence identity but do not result in an accurate model. Measures of statistical significance do not suffer from this normalization problem because the alignment scores are corrected for the length of the aligned segment before the significance is computed [37, 38].

9. After a list of all related protein structures and their alignments with the target sequence has been obtained, template structures are usually prioritized depending on the purpose of the comparative model. Template structures may be chosen based purely on the target–template sequence identity or a combination of several other criteria, such as the experimental accuracy of the structures (resolution of X-ray structures, number of restraints per residue for NMR structures), conservation of active-site residues, holo-structures that have bound ligands of interest, and prior biological information that pertains to the solvent, pH, and quaternary contacts.

10. Although fold assignment and sequence–structure alignment are logically two distinct steps in the process of comparative modeling, in practice almost all fold assignment methods also provide sequence–structure alignments. In the past, fold assignment methods were optimized for better sensitivity in detecting remotely related homologs, often at the cost of alignment accuracy. However, recent methods simultaneously optimize both the sensitivity and alignment accuracy. For the sake of clarity, however, they are still considered as separate steps in the current chapter.

11. Most alignment methods use either the local or the global dynamic programming algorithms to derive the optimal alignment between two or more sequences and/or structures. The methods, however, vary in terms of the scoring function that is being optimized. The differences are usually in the form of the gap penalty function (linear, affine, or variable) [14], the substitution matrix used to score the aligned residues (20×20 matrices derived from alignments with a given sequence identity, those derived from structural alignments, and those incorporating the structural environment of the residues) [39], or combinations of both [40–43]. There does not yet exist a single universal scoring function that guarantees the most accurate alignment for all situations. Above 30–40 % sequence identity, alignments produced by almost all methods are similar. However, in the twilight and midnight zones of sequence

identity, models based on the alignments of different methods tend to have significant variations in accuracy. Improving the performance and accuracy of methods in this regime remains one of the main tasks of comparative modeling [44, 45].

12. To generate each model, MODELLER takes a starting structure, which is simply the target sequence threaded onto the template backbone, adds some randomization to the coordinates, and then optimizes it by searching for the minimum of its scoring function. Since finding the global minimum of the scoring function is not guaranteed, it is usually recommended to repeat the procedure multiple times to generate an ensemble of models; the randomization is necessary otherwise, the same model would be generated each time. Computing multiple models is particularly important when the sequence–structure alignment contains different templates with many insertions and/or deletions. Calculating multiple models allows for better sampling of the different template segments and the conformations of the unaligned regions. The best scoring model among these multiple models is generally more accurate than the first model produced.
13. The MODELLER objective function is a measure of how well the model satisfies the input spatial restraints. Lower values of the objective function indicate a better fit with the input data and, thus, models that are likely to be more accurate [12].
14. The DOPE score [18] is an atomic distance-dependent statistical potential based on a physical reference state that accounts for the finite size and spherical shape of proteins. The reference state assumes that a protein chain consists of non-interacting atoms in a homogeneous sphere of equivalent radius to that of the corresponding protein. The DOPE potential was derived by comparing the distance statistics from a non-redundant PDB subset of 1,472 high-resolution protein structures with the distance distribution function of the reference state. By default, the DOPE score is not included in the model building routine and thus can be used as an independent assessment of the accuracy of the output models. The DOPE score assigns a score for a model by considering the positions of all non-hydrogen atoms, with lower scores predicting more accurate models. Since DOPE is a pseudo-energy dependent on the composition and size of the system, DOPE scores are only directly comparable for models with the same set of atoms (so can, for example, be used to rank multiple models of the same protein, but cannot be used without additional approximations to compare models of a protein and its mutant). The normalized DOPE (or z-DOPE) score, however, is a *z* score that relates the DOPE score of the model to the average observed DOPE

score for "reference" protein structures of similar size [25]. Negative normalized DOPE scores of –1 or below are likely to correspond to models with the correct fold.

15. Different measures to predict errors in a protein structure perform best at different levels of resolution. For instance, physics-based force fields may be helpful at identifying the best model when all models are very close to the native state (<1.5 Å RMSD, corresponding to ~85 % target–template sequence identity). In contrast, coarse-grained scores such as atomic distance statistical potentials have been shown to have the greatest ability to differentiate models in the ~3 Å Cα RMSD range. Tests show that such scores are often able to identify a model within 0.5 Å Cα RMSD of the most accurate model produced [46]. When multiple models are built, the DOPE score generally selects a more accurate model than the MODELLER objective function.

16. Segments of the target sequence that have no equivalent region in the template structure (i.e., insertions or loops) are among the most difficult regions to model [11, 47–49]. This difficulty is compounded when the target and template are distantly related, with errors in the alignment leading to incorrect positions of the insertions and distortions in the loop environment. Using alignment methods that incorporate structural information can often correct such errors [14]. Once a reliable alignment is obtained, various modeling protocols can predict the loop conformation, for insertions of less than approximately ten residues long [11, 47, 50, 51].

17. As a consequence of sequence divergence, the main-chain conformation of a protein can change, even if the overall fold remains the same. Therefore, it is possible that in some correctly aligned segments of a model, the template is locally different (<3 Å) from the target, resulting in errors in that region. The structural differences are sometimes not due to differences in sequence but are a consequence of artifacts in structure determination or structure determination in different environments (e.g., packing of subunits in a crystal and ligands). The simultaneous use of several templates can minimize this kind of error [52, 53].

18. It is particularly important to generate the best alignment of the structures to minimize conflicting information (e.g., one template suggesting that two Cα atoms in the target are close and another suggesting they are widely separated). SALIGN [15] uses both sequence- and structure-dependent features to align multiple structures. It employs an iterative procedure to determine the input parameters that maximize the structural overlap of the generated alignment.

Acknowledgments

We are grateful to all members of our research group. We also acknowledge support from National Institutes of Health (U54 GM094625) as well as computing hardware support from Ron Conway, Mike Homer, Hewlett-Packard, NetApp, IBM, and Intel.

References

1. Baker D, Sali A (2001) Protein structure prediction and structural genomics. Science 294(5540):93–96
2. Schwede T, Sali A, Honig B et al (2009) Outcome of a workshop on applications of protein models in biomedical research. Structure 17(2):151–159
3. Zhang Y (2008) Progress and challenges in protein structure prediction. Curr Opin Struct Biol 18(3):342–348
4. Marti-Renom MA, Stuart AC, Fiser A et al (2000) Comparative protein structure modeling of genes and genomes. Annu Rev Biophys Biomol Struct 29:291–325
5. Eswar N, Sali A (2009) Protein structure modeling. In: Sussman JL, Spadon P (eds) From molecules to medicine, structure of biological macromolecules and its relevance in combating new diseases and bioterrorism, NATO science for peace and security series: a—chemistry and biology. Springer, Dordrecht, The Netherlands, pp 139–151
6. Ginalski K (2006) Comparative modeling for protein structure prediction. Curr Opin Struct Biol 16(2):172–177
7. Das R, Baker D (2008) Macromolecular modeling with rosetta. Annu Rev Biochem 77: 363–382
8. Zhang Y, Skolnick J (2004) Automated structure prediction of weakly homologous proteins on a genomic scale. Proc Natl Acad Sci USA 101(20):7594–7599
9. Simons KT, Bonneau R, Ruczinski I, et al (1999) Ab initio protein structure prediction of CASP III targets using ROSETTA. Proteins (Suppl 3):171–176
10. Pieper U, Webb BM, Barkan DT et al (2011) ModBase, a database of annotated comparative protein structure models, and associated resources. Nucleic Acids Res 39:465–474
11. Fiser A, Do RKG, Sali A (2000) Modeling of loops in protein structures. Protein Sci 9(9): 1753–1773
12. Sali A, Blundell TL (1993) Comparative protein modelling by satisfaction of spatial restraints. J Mol Biol 234(3):779–815
13. Marti-Renom MA, Madhusudhan MS, Sali A (2004) Alignment of protein sequences by their profiles. Protein Sci 13(4):1071–1087
14. Madhusudhan MS, Marti-Renom MA, Sanchez R et al (2006) Variable gap penalty for protein sequence-structure alignment. Protein Eng Des Sel 19(3):129–133
15. Madhusudhan MS, Webb BM, Marti-Renom MA et al (2009) Alignment of multiple protein structures based on sequence and structure features. Protein Eng Des Sel 22:569–574
16. Brooks BR, Brooks CL 3rd, Mackerell AD Jr et al (2009) CHARMM: the biomolecular simulation program. J Comput Chem 30(10): 1545–1614
17. Sali A, Overington JP (1994) Derivation of rules for comparative protein modeling from a database of protein structure alignments. Protein Sci 3(9):1582–1596
18. Shen MY, Sali A (2006) Statistical potential for assessment and prediction of protein structures. Protein Sci 15(11):2507–2524
19. Wu G, Fiser A, ter Kuile B et al (1999) Convergent evolution of Trichomonas vaginalis lactate dehydrogenase from malate dehydrogenase. Proc Natl Acad Sci USA 96(11): 6285–6290
20. Berman HM, Westbrook J, Feng Z et al (2000) The protein data bank. Nucleic Acids Res 28(1):235–242
21. Smith TF, Waterman MS (1981) Identification of common molecular subsequences. J Mol Biol 147(1):195–197
22. Needleman SB, Wunsch CD (1970) A general method applicable to the search for similarities in the amino acid sequence of two proteins. J Mol Biol 48(3):443–453
23. John B, Sali A (2003) Comparative protein structure modeling by iterative alignment, model building and model assessment. Nucleic Acids Res 31(14):3982–3992
24. Melo F, Sanchez R, Sali A (2002) Statistical potentials for fold assessment. Protein Sci 11(2):430–448
25. Eramian D, Eswar N, Shen M et al (2008) How well can the accuracy of comparative

protein structure models be predicted? Protein Sci 17(11):1881–1893
26. Vajda S, Kozakov D (2009) Convergence and combination of methods in protein–protein docking. Curr Opin Struct Biol 19(2):164–170
27. Lensink MF, Wodak SJ (2010) Docking and scoring protein interactions: CAPRI 2009. Proteins 78(15):3073–3084
28. Alber F, Forster F, Korkin D et al (2008) Integrating diverse data for structure determination of macromolecular assemblies. Annu Rev Biochem 77:443–477
29. Russel D, Lasker K, Webb B et al (2012) Putting the pieces together: integrative structure determination of macromolecular assemblies. PLoS Biol 10(1):e1001244
30. Robinson C, Sali A, Baumeister W (2007) The molecular sociology of the cell. Nature 450(7172):973–982
31. Ward A, Sali A, Wilson I (2013) Structural biology unleashed. Science 339:913–915
32. Lasker K, Sali A, Wolfson HJ (2010) Determining macromolecular assembly structures by molecular docking and fitting into an electron density map. Proteins Struct Funct Bioinform 78:3205–3211
33. Tjioe E, Lasker K, Webb B et al (2011) MultiFit: a web server for fitting multiple protein structures into their electron microscopy density map. Nucleic Acids Res 39:167–170
34. Schneidman-Duhovny D, Hammel M, Sali A (2011) Macromolecular docking restrained by a small angle X-ray scattering profile. J Struct Biol 3:461–471
35. Rost B (1999) Twilight zone of protein sequence alignments. Protein Eng 12(2): 85–94
36. May AC (2004) Percent sequence identity; the need to be explicit. Structure 12(5):737–738
37. Altschul SF, Madden TL, Schaffer AA et al (1997) Gapped BLAST and PSI-BLAST: a new generation of protein database search programs. Nucleic Acids Res 25(17):3389–3402
38. Pearson WR (1998) Empirical statistical estimates for sequence similarity searches. J Mol Biol 276(1):71–84
39. Henikoff S, Henikoff JG (1992) Amino acid substitution matrices from protein blocks. Proc Natl Acad Sci USA 89(22):10915–10919
40. Zhou H, Zhou Y (2005) Fold recognition by combining sequence profiles derived from evolution and from depth-dependent structural alignment of fragments. Proteins 58(2): 321–328
41. McGuffin LJ, Jones DT (2003) Improvement of the GenTHREADER method for genomic fold recognition. Bioinformatics 19(7): 874–881
42. Karchin R, Cline M, Mandel-Gutfreund Y et al (2003) Hidden Markov models that use predicted local structure for fold recognition: alphabets of backbone geometry. Proteins 51(4):504–514
43. Shi J, Blundell TL, Mizuguchi K (2001) FUGUE: sequence-structure homology recognition using environment-specific substitution tables and structure-dependent gap penalties. J Mol Biol 310(1):243–257
44. Dunbrack RL Jr (2006) Sequence comparison and protein structure prediction. Curr Opin Struct Biol 16(3):374–384
45. Xiang Z (2006) Advances in homology protein structure modeling. Curr Protein Pept Sci 7(3):217–227
46. Eramian D, Shen M, Devos D et al (2006) A composite score for predicting errors in protein structure models. Protein Sci 15(7):1653–1666
47. Jacobson MP, Pincus DL, Rapp CS et al (2004) A hierarchical approach to all-atom protein loop prediction. Proteins 55(2):351–367
48. Zhao S, Zhu K, Li J et al (2011) Progress in super long loop prediction. Proteins 79(10): 2920–2935
49. Fernandez-Fuentes N, Oliva B, Fiser A (2006) A supersecondary structure library and search algorithm for modeling loops in protein structures. Nucleic Acids Res 34(7):2085–2097
50. van Vlijmen HW, Karplus M (1997) PDB-based protein loop prediction: parameters for selection and methods for optimization. J Mol Biol 267(4):975–1001
51. Coutsias EA, Seok C, Jacobson MP et al (2004) A kinematic view of loop closure. J Comput Chem 25(4):510–528
52. Sanchez R, Sali A (1997) Evaluation of comparative protein structure modeling by MODELLER-3. Proteins (Suppl 1):50–58
53. Srinivasan N, Blundell TL (1993) An evaluation of the performance of an automated procedure for comparative modelling of protein tertiary structure. Protein Eng 6(5):501–512
54. Sanchez R, Sali A (1998) Large-scale protein structure modeling of the Saccharomyces cerevisiae genome. Proc Natl Acad Sci USA 95(23):13597–13602
55. Chothia C, Lesk AM (1986) The relation between the divergence of sequence and structure in proteins. EMBO J 5(4):823–826

Chapter 2

RaptorX server: A Resource for Template-Based Protein Structure Modeling

Morten Källberg, Gohar Margaryan, Sheng Wang, Jianzhu Ma, and Jinbo Xu

Abstract

Assigning functional properties to a newly discovered protein is a key challenge in modern biology. To this end, computational modeling of the three-dimensional atomic arrangement of the amino acid chain is often crucial in determining the role of the protein in biological processes. We present a community-wide web-based protocol, RaptorX server (http://raptorx.uchicago.edu), for automated protein secondary structure prediction, template-based tertiary structure modeling, and probabilistic alignment sampling.

Given a target sequence, RaptorX server is able to detect even remotely related template sequences by means of a novel nonlinear context-specific alignment potential and probabilistic consistency algorithm. Using the protocol presented here it is thus possible to obtain high-quality structural models for many target protein sequences when only distantly related protein domains have experimentally solved structures. At present, RaptorX server can perform secondary and tertiary structure prediction of a 200 amino acid target sequence in approximately 30 min.

Key words Protein structure prediction, Homology modeling, Protein threading, Secondary structure prediction, Model quality assessment

1 Introduction

The advent of high-throughput procedures capable of identifying the entities making up cellular proteomes [1, 2] is one of the milestone accomplishments of recent decades. The availability of these high-dimensional datasets does, however, present us with the challenge of efficiently determining the functional role of the expressed protein entities. The biological activity of a protein domain, such as enzymatic catalysis [3] or signaling transduction [4], is often highly related to the three-dimensional arrangement of its amino acid chain. Structural models of newly discovered proteins are thus valuable in uncovering their biological function and can serve as an important stepping stone in generating hypotheses or suggesting

Daisuke Kihara (ed.), *Protein Structure Prediction*, Methods in Molecular Biology, vol. 1137, DOI 10.1007/978-1-4939-0366-5_2,

experiments to further explore their nature. While the Protein Data Bank (PDB) [5] provides experimentally determined structural data for a number of protein domains, the vast majority of protein sequences available in public databases currently do not have solved structures.

To this end template-based modeling methods can generate approximate models for a large number of sequences with relative ease if a closely related template domain sequence with solved structure is available. Current methods do, however, become unreliable when there are no homologs with solved structures in the PDB or when templates under consideration are distant homologs [6]. Template-based modeling is critically dependent on the quality of the target–template alignment. To better address cases where no close template exists, we studied and implemented a number of novel modeling strategies in our new software RaptorX server [7, 8]. RaptorX server takes into consideration the number of nonredundant homologs available for the target sequence and a template structure to assess the quality of information content in sequence profiles [9]. This allows us to optimize the modeling strategy specifically to the target. Second, RaptorX server uses conditional neural fields (CNF), a variant of conditional random fields (CRF), to integrate a variety of context-specific biological signals in a nonlinear probabilistic scoring function [10]. Finally, RaptorX server has also implemented a multiple-template threading (MTT) procedure [11], enabling the use of multiple templates to model a single-target sequence. Results from CASP9 and the recently concluded CASP10 competitions clearly demonstrate the value of the abovementioned innovations. RaptorX server ranked second being only outperformed by a server employing consensus analysis of results from multiple single methods and extensive post-threading refinement [12].

Aside from structure modeling, RaptorX server provides options for custom pairwise target–template alignments and single-target multiple-template alignments. Furthermore, RaptorX server utilizes a CNF [13]-based prediction protocol for determining the three-state secondary structure, eight-state secondary structure, and solvent accessibility distributions for each residue in the target sequence. RaptorX server also provides disorder prediction of an input protein sequence.

The secondary and tertiary structure models generated by RaptorX server can serve as starting points for further analysis in a number of diverse application areas. For example, the predicted 3D models can be used for binding site epitope prediction as well as in protein docking and protein–protein interaction studies [14, 15].

2 Materials

The following are necessary for the use of RaptorX server.

1. A personal computer connected to the Internet and a web browser with Java Script enabled: RaptorX server is compatible with three popular web browsers: Google Chrome, Firefox, and Internet Explorer. Nevertheless, the former two browsers may be slightly better than the third one in visualizing the prediction results.
2. The amino acid sequence(s) of the protein(s) of interest in FASTA format: The allowed characters in the sequence are the one-letter codes for the 20 standard amino acids. Spaces and line breaks in the sequence string are ignored and do not affect the prediction. To prevent a single sequence from occupying the server for a very long time, we currently limit the length of user-submitted sequences to 2,000 amino acids.

3 Methods

In this section we present two separate use cases of the RaptorX server. First, we cover the main use case of obtaining the secondary and tertiary structures of a target sequence. Second, we demonstrate the use of RaptorX server to generate alignments between the target sequence and user-specified template structures.

3.1 Modeling the Secondary and Tertiary Structures of a Target Sequence

1. In the web browser navigate to http://raptorx.uchicago.edu.
2. From the menu at the top of the page select "New job."
3. Use the tab menu to choose between "Alignment Job" and "Structure Prediction Job."
4. In the "Job Identification" section of the form provide a job name (defaults to "my job") and an e-mail address that will be used for notification upon job completion. The e-mail given also serves as the username for accessing results at a later date. Since RaptorX server does not require any user registration, it is important that a correct e-mail address is provided.
5. In the "Sequences" box, provide one or more FASTA-formatted sequences. These can be supplied by copy and pasting into the text box or by uploading a flat-text file with the data. The FASTA identifier is used to identify the individual sequence(s) when browsing prediction results; therefore, we recommend using descriptive sequence names. In the "Job Settings" section, choose if multiple-template modeling is to be used (recommended) and if you wish to do secondary, tertiary, or both secondary and tertiary structure modeling.

6. Press the submit button to queue the prediction job. The data entered in the form will be validated, and the user will be notified of any errors that need correction in a box appearing at the top of the page. If the submission is successful the user will be redirected to an overview page displaying pending and completed prediction jobs. It should be noted that the number of pending jobs allowed for one user is limited to 20.
7. In order to track pending and completed prediction jobs the user needs to be logged in to the server. If the login from a previous session has expired or the account needs to be accessed from a different machine than the one used for the initial submission, the user can supply the account e-mail in the login field on the RaptorX server front page. An e-mail message will be sent to the address containing a hyperlink to the overview page.
8. Select "My jobs" in the menu at the top of the page to display the job overview for the account. Here, the status of each prediction in the job is given along with overall information on the predictions being done for each submitted sequence. To track the job status in real time simply refresh the page, and the completion status of the prediction for each submitted sequence in a job will be updated (*see* **Note 1**).
9. Click on the structure labeling link in the job overview page to bring up a summary page similar to the one depicted in Fig. 1.
10. Structure labeling prediction is provided in four modes. The available results include three-class and eight-class secondary structures, disorder prediction, as well as three-state solvent accessibility. You can switch between the modes using the blue tab menu (*see Label 1* in Fig. 1). The three-class secondary structure prediction gives the distribution between the classes alpha-helix, extended strand in beta ladder, and loop/irregular. In addition to these, the eight-class prediction classes include residue in isolated beta-bridge, 3-helix (3/10 helix), 5-helix (π-helix), hydrogen-bonded turn (3, 4, or 5 turn), and Bend (*see* **Note 2**). Disorder prediction classifies residues as disorder or non-disorder, while solvent accessibility classes are buried, medium, and exposed.
11. For each residue a figure depicting the distribution of structure labeling classes is given, indicating the relative likelihood of a given residue belonging to each of these classes. The legend for the color coding of the states can be found in the column on the right-hand side of the page (*see Label 5* in Fig. 1). Hover over a residue to display the exact probability distribution of secondary structure classes in a pop-up box next to the residue (*see Label 2* in Fig. 1).
12. The right-hand column provides information on the status of the prediction job (*see Label 3* in Fig. 1); to download the

Fig. 1 Example of a secondary structure prediction result

prediction results for the sequence, including the full class distributions of the four secondary structure predictions, click the link labeled "Download" (*see Label 4* in Fig. 1).

13. Click on a 3D structure link in the job overview page to obtain a job summary similar to the one depicted in Fig. 2a, b (*see* **Note 3**).
14. In a structure prediction job, a protein structure is built for each of the (≤10) top-ranked alignments between the target and sequences from the template library. The rank of the candidate model is provided in the results overview (*see Label 1* in Fig. 2a), with the highest ranked model being selected as the default. Clicking the "View alternative models" button will bring up a menu from which the user can switch between models (*see Label 5* in Fig. 2a). For each model, the PDB code of the template along with the *p*-value and uGDT score of the alignment is given. If MTT is used, a model based on several templates will be available as well (*see Label 4* in Fig. 2a) (*see* **Note 4**).
15. The quality of the model is given by *p*-value, uGDT, and global distance test (GDT) (*see Labels 2 and 3* in Fig. 2a) of its

a

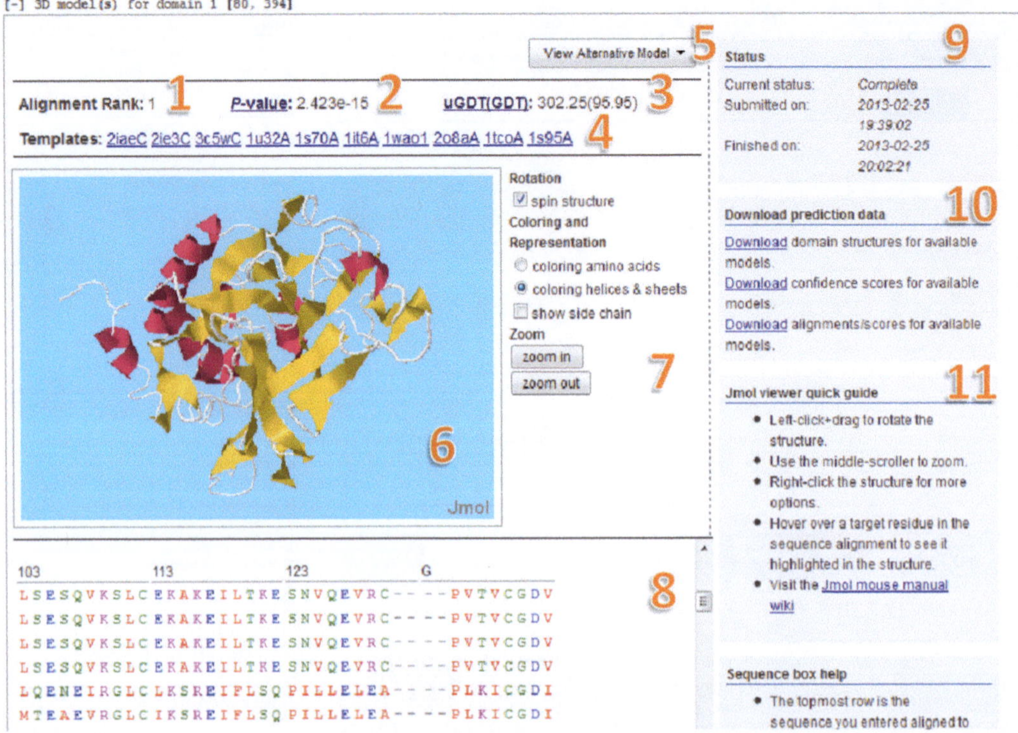

b

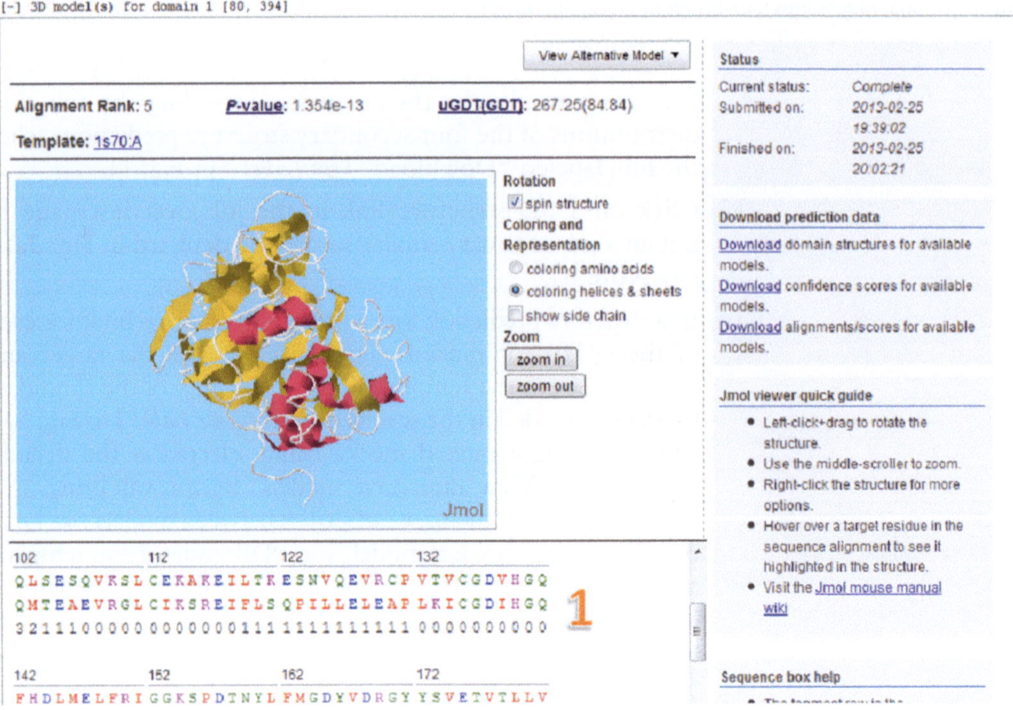

Fig. 2 (**a**) Example of a tertiary structure prediction result with multiple templates. (**b**) Example of a tertiary structure prediction result with a single template and local quality score

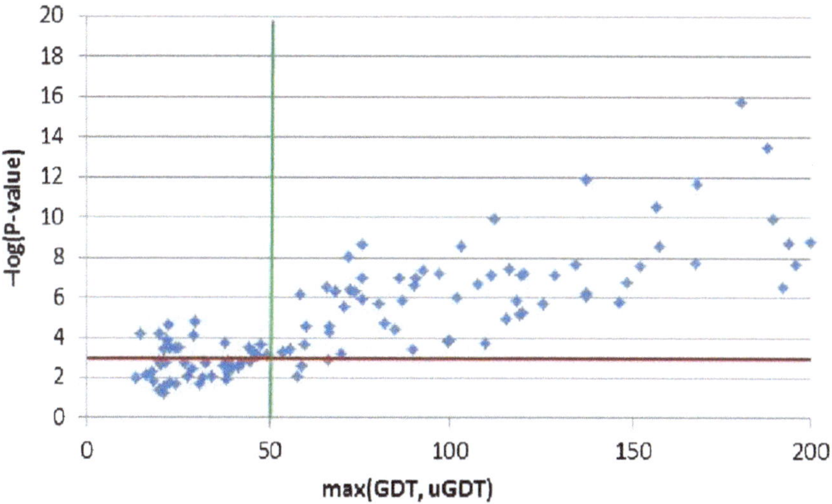

Fig. 3 The relationship between *p*-value and the model quality on the 123 CASP10 targets

alignment with the selected template. The uGDT is the unnormalized GDT score which is defined as $1 \times N(1) + 0.75 \times N(2) + 0.5 \times N(4) + 0.25 \times N(8)$, where $N(x)$ is the number of residues with the local RMSD smaller than xÅ. GDT is uGDT normalized by a protein domain length. GDT measures the quality of a model by comparing it with the native structure and has a value ranging from 0 to 100, indicating the worst and the best quality, respectively. As shown in Fig. 3, the *p*-value is a reliable indicator of model quality. When the *p*-value is small (i.e., $<10^{-5}$), the models have a uGDT or a GDT greater than or equal to 50. Even in the case of a *p*-value smaller than 10^{-4}, only three models have both uGDT and GDT less than 50. That is, the prediction from our threading method is reliable when the *p*-value is less than 10^{-4}. For each model, the PDB identifier for the template structure and the specific polypeptide chain from the PDB file used to build the currently selected model are displayed. Click the link to go to the structure record in the PDB (http://www.pdb.org) (*see Label 4* in Fig. 2a).

16. A graphic representation of the currently selected model is provided in the Jmol viewer. Use the mouse to rotate and zoom the structure. Right-clicking on the model will bring up a menu of further options for changing the visualization settings (*see Label 6* in Fig. 2a). To the right of the structure viewer a menu for controlling the representation of the currently selected model is available (*see Label 7* in Fig. 2a).

17. The alignment of the target and template sequences used for constructing the current model is displayed below the Jmol viewer. Each position in the alignment is color coded according

to the chemical nature of the residue. The color scheme used is the following: Red = Hydrophobic, Blue = Acidic, Magenta = Basic, and Green = Hydroxyl + Amine. RaptorX server also provides the predicted RMSD at each aligned position rounded to the nearest integer as indicators of reliability in the last row (*see Label 1* in Fig. 2b). Hover over the aligned residues to highlight the corresponding target residues in the Jmol viewer (*see Label 8* in Fig. 2a).

18. The column to the right provides an overview of the prediction job status (*see Label 9* in Fig. 2a). Click on the appropriate links to download the prediction results. Multiple download options are available: PDB files for the top-ranked models, the corresponding alignments with their local reliability scores, and the confidence scores such as the *p*-value, uGDT, and GDT mentioned above (*see Label 10* in Fig. 2a). Underneath the download links a third box with a brief user's guide for the Jmol viewer is given (*see Label 11* in Fig. 2a) followed by a guide on the sequence box.

3.2 Custom Template Alignment

19. Repeat **steps 1–5** from Subheading 3.1 above.
20. Indicate the structure(s) you wish the supplied sequence(s) from **step 5** to be aligned to. Enter the PDB ID in the text box, and select the desired structure from the drop-down menu that appears. Repeat to add additional structures to the list (*see* **Note 5**).
21. Under "Alignment options," check the types of alignments you wish to generate. The options available are "Optimal pairwise alignment" which returns the best possible pairwise alignment between the target sequence and the selected templates; "Probabilistic sampling" which returns a user-specified number of alternative alignments sampled according to the alignment probability distribution generated by the CNF model; or "Multiple template alignment" which returns a multiple-protein alignment between the selected templates and the input target sequence.
22. Click on an alignment job in the job overview to obtain a summary similar to the one depicted in Fig. 4.
23. In an alignment job, in addition to the optimal alignments between the target sequence and the provided template structures, a set of sampled alternative alignments may also be generated. To generate a sample alignment, check the "Probabilistic sample" box and indicate the number of samples desired.
24. Click on the alignment drop-down selection box to bring up a selection menu from which it is possible to switch between alternative alignments (*see Label 1* in Fig. 4). The alignment of the target and template sequences will be displayed after a selection is made, and the "Display" button is pressed.

Fig. 4 Example of a custom alignment result

Each position in the alignment is color coded according to the chemical nature of the residue as described in **step 17** (*see Labels 2 and 5* in Fig. 4).

25. The right-hand column provides information on the status of the job (*see Label 3* in Fig. 4). Click on the links to download the alignment results, including the set of alignments between the target sequence and all structures in the template library used (*see Label 4* in Fig. 4).

4 Notes

1. From time to time the user may not receive a response from the server after submitting several sequences to RaptorX server for 1 or 2 days. RaptorX server can usually process at least one of the submitted sequences within 24 h even when operating at a high load; however, exceptionally heavy loads may delay the response. Other possible reasons for delay include server maintenance or an incorrect e-mail address provided by the user. Click on the "contact" menu at the bottom of the RaptorX server web page, and send a message to the system administrator.
2. Sometimes the user may observe different probabilities for the same secondary structure class for a given residue in the three- and eight-state models. As an example, residue 13 is in

an alpha-helix with probability 17 % in the three-state model and 14 % in the eight-state model. As the two models give the distribution of secondary structure groups from two different class sets, the differences in the alpha-helix propensity between the two models could be due to other types of helices being possible in the eight-state model.

3. It should be noted that the prediction results are not expanded automatically when a results page is loaded. This is done to provide a better overview for the submitted sequence consisting of many domains. For any one submission there will be at least four entries in the result page including secondary and tertiary structure prediction, domain parsing, and disorder prediction. Clicking on any of them will display the relevant result.
4. Even if MTT is selected you may not see any MTT results in the drop-down menu. MTT is only deployed if our method predicts that a model based on several template structures is more accurate than the top-ranked single-template model. Should you still want to construct a multiple-template alignment, this can be accomplished through the custom alignment interface.
5. When looking up a template structure in the drop-down menu you may not always be able to find the desired PDB identifier. This is due to the template library used on the server being "non-redundant"; thus, several highly similar structures in the PDB are omitted, and only one representative structure is kept in the library. To resolve this problem, we supply a list of equivalent structures to identify the structure in the library equivalent to your desired template.

Acknowledgments

This work is supported by the National Institute of Health grant R01GM0897532, National Science Foundation DBI-0960390 and CAREER award, Alfred P. Sloan Fellowship, and TTIC summer intern program. The authors are grateful to the University of Chicago Beagle team, TeraGrid, and Canadian SHARCNet for their support of computational resources.

References

1. Aebersold R, Mann M (2003) Mass spectrometry-based proteomics. Nature 422(6928):198–207
2. Källberg M, Lu H (2010) An improved machine learning protocol for the identification of correct Sequest search results. BMC Bioinformatics 11:591
3. Bairoch A (2000) The ENZYME database in 2000. Nucleic Acids Res 28(1):304–305
4. Hannum G et al (2009) Genome-wide association data reveal a global map of genetic interactions among protein complexes. PLoS Genet 5(12):e1000782
5. Berman HM et al (2000) The protein data bank. Nucleic Acids Res 28(1):235–242
6. Baker D, Sali A (2001) Protein structure prediction and structural genomics. Science 294(5540):93–96

7. Peng J, Xu J (2011) RaptorX: exploiting structure information for protein alignment by statistical inference. Proteins 79(Suppl 10):161–171
8. Kallberg M et al (2012) Template-based protein structure modeling using the RaptorX web server. Nat Protoc 7(8):1511–1522
9. Peng J, Xu J (2010) Low-homology protein threading. Bioinformatics 26(12):i294–i300
10. Peng J, Xu J (2009) Boosting protein threading accuracy. Lect Notes Comput Sci 5541:31
11. Peng J, Xu J (2011) A multiple-template approach to protein threading. Proteins 79(6): 1930–1939
12. Roy A, Kucukural A, Zhang Y (2010) I-TASSER: a unified platform for automated protein structure and function prediction. Nat Protoc 5(4):725–738
13. Peng J, Bo L, Xu J (2009) Conditional neural fields. In: Bengio Y, Schuurmans D, Lafferty J, Williams CKI, Culotta A (eds) Advances in neural information processing systems, vol 22. p 1419–1427
14. Singh R et al (2010) Struct2Net: a web service to predict protein–protein interactions using a structure-based approach. Nucleic Acids Res 38:W508–W515
15. Singh R, Xu J, Berger B (2006) Struct2net: integrating structure into protein–protein interaction prediction. Pac Symp Biocomput 2006:403–414

Chapter 3

The MULTICOM Protein Tertiary Structure Prediction System

Jilong Li, Debswapna Bhattacharya, Renzhi Cao, Badri Adhikari, Xin Deng, Jesse Eickholt, and Jianlin Cheng

Abstract

With the expansion of genomics and proteomics data aided by the rapid progress of next-generation sequencing technologies, computational prediction of protein three-dimensional structure is an essential part of modern structural genomics initiatives. Prediction of protein structure through understanding of the theories behind protein sequence–structure relationship, however, remains one of the most challenging problems in contemporary life sciences. Here, we describe MULTICOM, a multi-level combination technique, intended to predict moderate- to high-resolution structure of a protein through a novel approach of combining multiple sources of complementary information derived from the experimentally solved protein structures in the Protein Data Bank. The MULTICOM web server is freely available at http://sysbio.rnet.missouri.edu/multicom_toolbox/.

Key words Protein tertiary structure, Template recognition, Multiple template combination, Protein structure prediction, Structure quality evaluation, Structure quality enhancement

1 Introduction

The past few decades have witnessed an explosive growth in genomics and proteomics data. With the advancement of high-throughput genome sequencing technologies, the total number of gene and protein sequences is increasing exponentially. Therefore, in this genomic era, one vital goal for life scientists is to acquire knowledge from this vast repository of resources for better drug design and disease prevention strategies. Proteins fold into a three-dimensional structure, called tertiary structure, in order to carry out necessary biological functions, and therefore a high-resolution tertiary structure of a protein is the key to understanding and manipulating its biochemical and cellular functions. However, the rate of protein structure determination by experimental techniques (e.g., X-ray crystallography or NMR spectroscopy) lags far behind the rate of acquisition of new protein sequences primarily due to

Daisuke Kihara (ed.), *Protein Structure Prediction*, Methods in Molecular Biology, vol. 1137,
DOI 10.1007/978-1-4939-0366-5_3, © Springer Science+Business Media New York 2014

the time-consuming and expensive nature of the experimental methods. Therefore, the gap between known protein sequences and structure will continue to widen in the future making it impossible to experimentally solve the structures for all proteins. Consequently, less expensive and time-efficient computer-assisted prediction of protein tertiary structures is becoming increasingly popular.

Around 50 years ago, Anfinsen discovered the fact that all of the information necessary for RNase A to fold into its native structure is contained in its amino acid sequence, suggesting that the structure of a protein could be derived uniquely from its sequence alone [1]. Subsequently, interpretation of the sequence–structure relationships in proteins has become an active area of research in the field of biological sciences. As soon as the experimental structures of the first few proteins were made available, it became clear that evolutionarily related (homologous) proteins tend to retain the same overall three-dimensional fold (i.e., the arrangement and association of structural fragments) while accumulating some divergent mutations [2]. Moreover, despite being strongly correlated, structural divergence is much slower than sequence divergence [3]. These two important findings gave birth to one doctrine in protein structure prediction (also known as protein modeling) called homology modeling or comparative modeling (CM) [4]. Traditionally, this technique attempts to map the sequence of one protein (a target) to the sequence of another protein with a known structure (a template) to deduce the overall fold of the target and subsequently alter the target structure according to its sequence divergence with respect to the template. This approach is also commonly known as template-based modeling (TBM) and is one of the most widely used techniques in computational protein structure prediction. Intuitively, the success of TBM depends largely on the availability and ability to identify suitable templates for the target as well as the sequence similarity between the target and template. The accuracy is usually low when only a relatively distant homologous template is available for the target. Promisingly, constant efforts have been made by the community in the last decade, resulting in continual improvement of the accuracy of computationally based structure prediction.

With the aim of an objective assessment of the improvement in state-of-the-art methods for protein structure prediction, Moult and co-workers organized the biennial community-wide experiment called critical assessment of techniques for protein structure prediction (CASP) [5]. It was clear from the assessment of the CASP blind experiment that the accuracy of computational protein structure can be improved by combining information from multiple templates instead of relying on a single template [6–8]. This concept is at the heart of the MULTICOM protein structure prediction system [9]. MULTICOM essentially is a robust framework which aligns the target protein with multiple complementary templates and attempts to enhance the accuracy of structure prediction using a novel model combination approach followed by quality assessment techniques [10, 11] to refine the alternative models with the goal of selecting the best structure. MULTICOM officially made its debut in CASP8 [12], and the assessment of the results demonstrates the effectiveness of the method across diverse target difficulties (i.e., for easy cases where a suitable template can be identified to hard cases where only distantly homologous templates are available). With its consistent success during the CASP9 experiment, MULTICOM has been acknowledged by the community as one of the "best public CASP-certified protein structure prediction servers" (http://predictioncenter.org/index.cgi?page=links).

In the subsequent sections, we attempt to provide a thorough and comprehensive overview of the MULTICOM protein structure prediction suite. Subheading 2 (Materials) describes the input data, step-by-step instructions on how to use the MULTICOM web interface in order to generate the tertiary structure of a protein, and how to interpret the results. In Subheading 3 (Methods), we provide methodologies used to develop the multi-level combination pipeline used in MULTICOM. Two representative examples have been furnished in Subheading 4 (Case Studies) for users which describe the typical use of the system and the way to analyze the output. Subheading 6 (Notes) covers some beneficial tips to aid the users of MULTICOM on how to use the system seamlessly and resolve any potential issues during the execution of the pipeline or analysis of the results.

2 Materials

2.1 Input

The input for the MULTICOM web server is the single-lettered amino acid sequence of the protein whose tertiary structure is to be predicted. The web server also needs a target name and e-mail address along with the amino acid sequence. The target name uniquely identifies the job, which is helpful when there is more than one job being submitted. The e-mail address is where the server sends the predicted model once the prediction is complete.

2.2 Usage

Predicting a protein's structure using MULTICOM is a two-step process. The first step is to submit the amino acid sequence to the server and then wait for the results. The second step begins after the MULTICOM web server sends an e-mail with the predicted structure as an attachment. The attached structure file is a standard protein data bank (pdb) file and can be visualized, analyzed, or evaluated using any available tools.

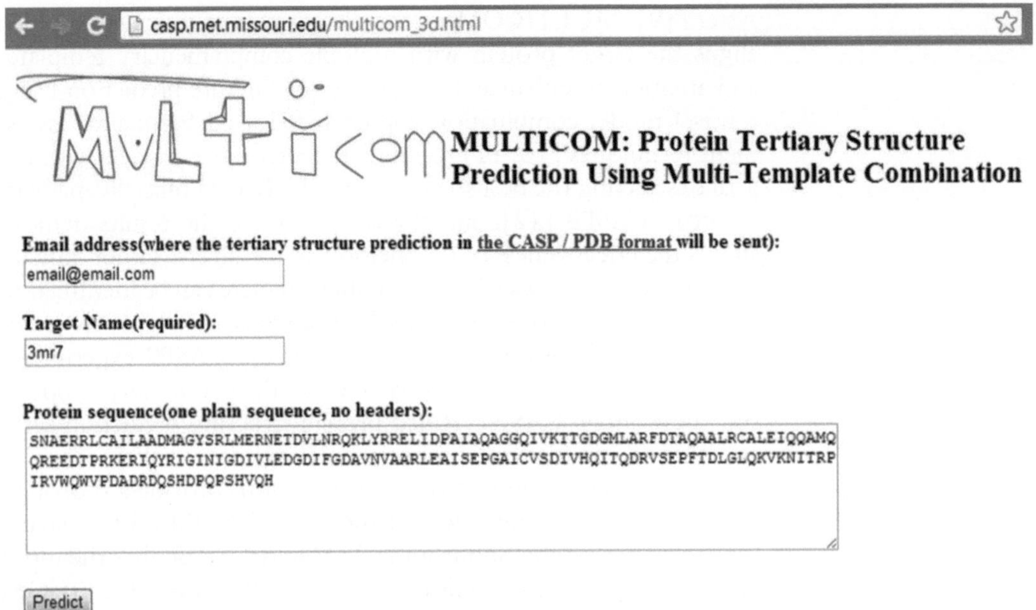

Fig. 1 The MULTICOM web server input page being filled with the sequence of chain A of a protein with PDB ID 3MR7. The input sequence is text wrapped in the text area and does not contain any white space characters

2.2.1 Step 1: Submit the Sequence

The input sequence of amino acids should not contain any letters or characters other than the 20 standard amino acid symbols. Any special characters such as *, $, and & should be removed from the sequence. White space characters including space, newline, tab, and carriage return should also be removed from the sequence. Once the e-mail address, target name, and sequence fields are filled, clicking on the predict button displays a status page. Figure 1 shows an example input for chain A of the protein with PDB ID 3MR7. All data in the input fields, including the e-mail address, needs to be verified before clicking on the predict button.

2.2.2 Step 2: Download the Prediction

Once the server completes the prediction, the results are sent to the corresponding e-mail address. The e-mail sent by the MULTICOM web server contains two attachments: model.pdb and align.pir. The PDB codes of the template sequences along with their alignment score are also included in the e-mail body as a list.

2.3 Output

The pdb file attached is the standard pdb file that has the *x*, *y*, and *z* coordinates of each atom in the protein and is in standard CASP format (http://predictioncenter.org/casp8/index.cgi?page=format). The pir file attached is a multiple sequence alignment file that shows sequence alignment of the input sequence with the templates found during the prediction process and is used to generate the predicted structure. The pdb file can be visualized using any viewer tools such as Chimera [13], PyMOL [14], Rasmol [15], and Jmol [16].

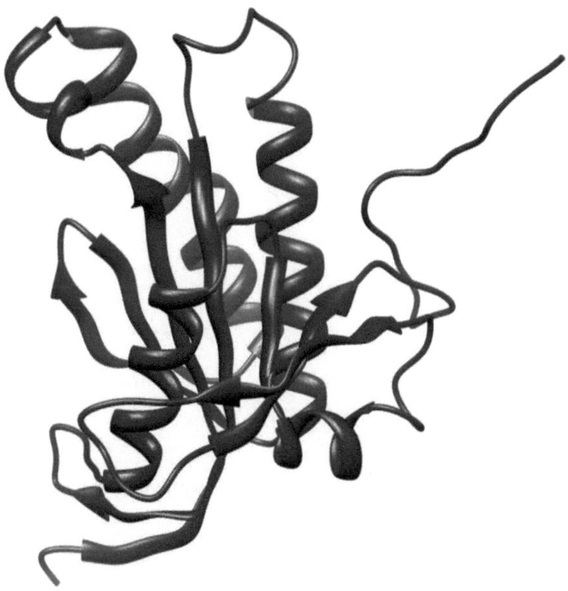

Fig. 2 The MULTICOM web server's prediction for chain A of a protein with PDB ID 3MR7 visualized using PyMOL

Figure 2 shows an example of visualizing the model.pdb file predicted for chain A of a protein with PDB ID 3MR7. In case the native structure is also available, tools like TM-score [17] may be used to evaluate the prediction. Additionally, the alignment file may be analyzed for alignment information in order to understand the contribution of each template to the predicted model.

2.4 Availability

The MULTICOM web server is freely accessible at http://casp.rnet.missouri.edu/multicom_3d.html which is in the MULTICOM toolbox (http://sysbio.rnet.missouri.edu/multicom_toolbox/). Prediction time depends on factors including server load, length of the input sequence, and difficulty of the query (i.e., whether or not good templates can be found).

3 Methods

As shown in Fig. 3, there are five steps in the MULTICOM protein structure prediction system [9, 18]. The first step generates a number of templates and their sequence alignments for an input query sequence. The second step generates a number of query-template alignments. The third step creates several structures (also called protein models) for the query. The fourth step evaluates the quality of the generated models. The last step improves the quality of the generated models. Finally, the system outputs the predicted model with the best quality.

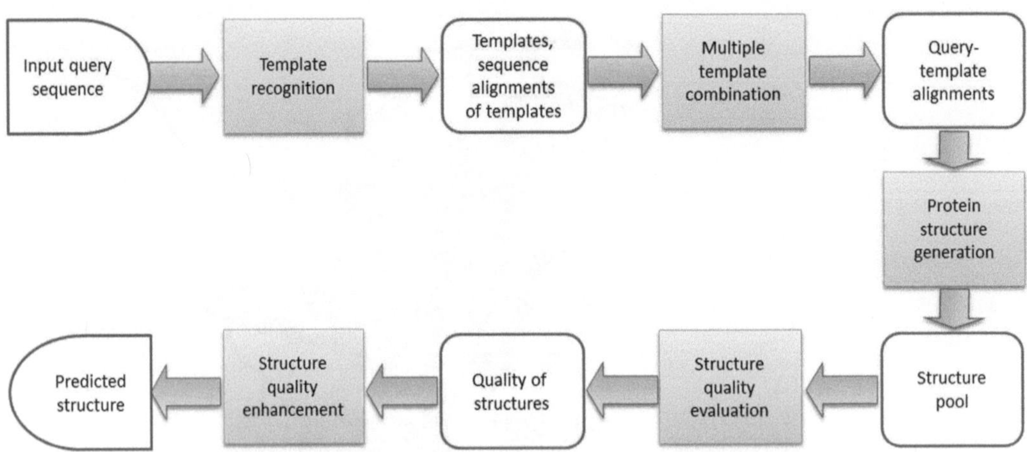

Fig. 3 The MULTICOM protein tertiary structure prediction system

3.1 Template Recognition

Template recognition needs a template library in order to identify the templates for the query sequence. In this system, the template library has been constructed based on the PDB [19]. The template library includes information such as template sequence, template structure, secondary structure, solvent accessibility, and template sequence profiles.

In this step, sequences homologous to the query are first found by searching the query sequence against the non-redundant protein sequence database via PSI-BLAST [20]. The query and its homologous sequences are then searched against the template library by different search tools [20–27] in order to find a number of templates with information about the structure of the query. A number of templates with low *e*-values are generated after these searches, along with local alignments between the query and its templates (*see* **Note 1**). The top-ranked templates and their query-template alignments for each tool are saved separately. A consensus list of the top-ranked templates is also generated according to the number of times it is identified by each search tool.

3.2 Multiple Template Combination

This step integrates multiple template structures coming from the previous step and generates a number of combined query-template alignments. This is done because multiple structurally similar templates may provide more accurate structural information for the query than a single template [6]. Three multiple template combination methods are used in this step.

The first method creates a combined query-template alignment based on the query-template alignments generated by each search tool. The combined query-template alignment contains the best query-template alignment and some other query-template alignments that have similar *e*-values with the best alignment. The aligned regions of all alignments have consistent structures (*see* **Note 2**).

The second method creates a combined query-template alignment based on the consensus list of templates. For each template, TM-Align [28] is used to align it with all other templates and the aligned regions are used to generate the multiple sequence alignment of this template. Then the multiple sequence alignment tool is used to align the multiple sequence alignments of all templates and that of the query to get the combined query-template alignment.

The third method uses three kinds of query-template alignments generated by PSI-BLAST [20], HHSearch [25], and SPEM [29] separately. This method combines these alignments for one query in this order: the PSI-BLAST local alignment, HHSearch alignment, and SPEM global alignment.

3.3 Protein Structure Generation

This step first checks the templates identified by the previous steps. If there are one or more templates which can cover the whole query or most of the query with very short unaligned regions (*see* **Notes 2** and **3**), the TBM tool Modeller [30] is used to generate a number of models. If there are no homologous templates or only one template covering a part of the query, a recursive protein modeling method [31] is used to generate the models. This method first uses the TBM tool Modeller [30] to model the regions which are aligned and covered very well by templates. We call these regions certain regions, while the unaligned regions are termed uncertain regions. A variant of Rosetta [31, 32] is used to construct other uncertain regions. Depending on the amount of template information available, the method may use only the TBM method or template-free modeling method or combine TBM method and template-free modeling method to generate a structure for the query. The final product of this step is a model pool for the query.

3.4 Structure Quality Evaluation

This step evaluates the quality of each model without knowing the native structure. In order to evaluate the quality of each model and identify the more accurate models, three structure quality evaluation methods are used. The first method (ModelEvaluator [33]) provides each model with an absolute quality score based on the features of that model (*see* **Note 4**). The secondary structure, solvent accessibility, contact map, and beta-sheet topology of the model can be parsed from the model directly, and they also can be predicted from the target sequence [34–36]. For each of them, we use the difference between that parsed from the model and that predicted from the target sequence as a feature. The second approach uses the structure alignment tool TM-score [17] to calculate the similarity score between the model and all other models in the model pool and then uses the average similarity score as the quality score of this model (*see* **Note 5**). The third method tries to combine the first two approaches. It selects the top models based on the quality score using the first method as the reference model set. Each model is compared with all models in the reference model set, and the average similarity score is used as the quality score. The local quality score of each residue is also calculated in this step. This is accomplished by aligning a model with each model in the reference model set. The distance between each residue in this model and its counterpart in a reference model in the reference set is calculated separately as a local quality score. Finally, the local quality score of each residue is the average distance of this residue and all of its counterparts.

3.5 Structure Quality Enhancement

In this step, the top-ranked models based on the structure quality evaluation are searched against the model pool to check if there exist other similar models (*see* **Note 6**). If there are some similar models, this step combines the top-ranked models with the similar models. Otherwise, very similar local regions of other models are combined with the top-ranked models. This model combination can usually get better models than the original top-ranked models. Moreover, the local quality score is also used for the structure quality enhancement. The regions with very poor local quality scores are resampled by a variant of Rosetta [31, 32] which constrains the local region modeling without changing other regions. The final prediction of this system is the best refined model.

4 Case Studies

As case studies, the MULTICOM web server was used to predict tertiary structure of the first chains (chain A) of two proteins: adenylate/guanylate cyclase/hydrolase from *Silicibacter pomeroyi* and diguanylate cyclase from *Pelobacter carbinolicus*. These proteins were also listed as prediction targets in CASP9 with target id as T0520 (http://predictioncenter.org/casp9/target.cgi?id=21&view=all) and T0634 (http://predictioncenter.org/casp9/target.cgi?id=178&view=all), respectively. These two protein sequences were supplied to the MULTICOM web server. The predictions were visualized using PyMOL and evaluated using TM-score and RMSD (average root mean square distance between the corresponding atoms) (*see* **Note 7**). The case studies show that the predicted structures are highly accurate with TM-score value of 0.9454 for target T0520 and 0.8547 for target T0634 and an RMSD value of 0.581 for T0520 and 1.257 for T0634. MULTICOM was ranked among the top ten predictors for both of these targets.

4.1 Case Study I

To predict the tertiary structure of adenylate/guanylate cyclase/hydrolase (from *Silicibacter pomeroyi*), its corresponding fasta sequence file was downloaded from PDB [19]. The PDB ID for

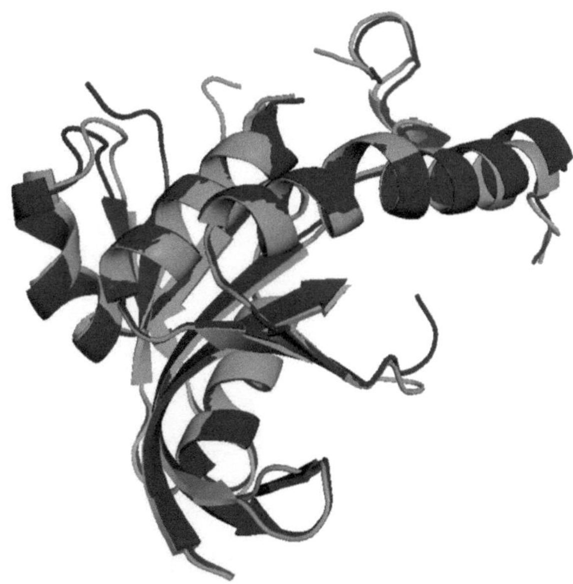

Fig. 4 Filtered native structure (shown in *green* color) and MULTICOM-predicted filtered structure (shown in *blue* color) superimposed using PyMOL for protein adenylate/guanylate cyclase/hydrolase

this protein is 3MR7, and the fasta sequence file is available at http://www.rcsb.org/pdb/files/fasta.txt?structureIdList=3MR7. The sequence for chain A was copied to a separate text file to remove newline characters. After removing newline characters, the whole sequence, 189 characters long, was now in a single line that begins with the residues SNAE and ends with residues HVQH. The sequence was then copied and supplied as input to the MULTICOM web server as shown in Fig. 1. The server took 17 min to complete the task. The predicted structure (model.pdb) was then visualized with PyMOL. To visually compare the predicted structure with the native structure, the native structure was downloaded from http://www.rcsb.org/pdb/files/3MR7.pdb. Before performing the comparison, the native structure and predicted structure both need to be filtered for two reasons: the native structure has three chains, and the predicted structure has only one; thus, there may be disordered regions in predicted or native structures. Finally, the filtered predicted structure and filtered native structure were both superimposed and visualized in PyMOL as shown in Fig. 4. Additionally, the predicted structures were evaluated using TM-score and RMSD (*see* **Note** 7). The TM-score value of 0.9454 and RMSD value of 0.581 show that the prediction is very accurate.

4.2 Case Study II

To predict the structure of diguanylate cyclase (from *Pelobacter carbinolicus*), steps similar to Case Study I were executed. The PDB ID for this protein is 3N53, and the fasta file was downloaded from

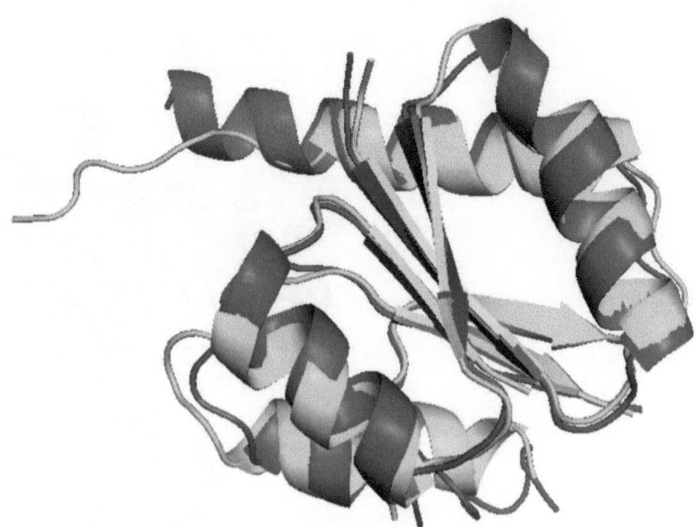

Fig. 5 Filtered native structure (shown in *pink* color) and MULTICOM-predicted filtered structure (shown in *yellow* color) superimposed using PyMOL for protein diguanylate cyclase

http://www.rcsb.org/pdb/files/fasta.txt?structureIdList=3N53. The 140-residue-long sequence starting with MSLK and ending with HHHH was supplied to the web server, and it took around half an hour for the prediction to complete. Similar to Case Study I, the filtered predicted structure and filtered native structure were obtained, superimposed, and visualized in PyMOL as shown in Fig. 5. For this target as well, a high TM-score value of 0.8547 and RMSD value of 1.257 imply an accurate prediction.

5 Conclusion

Given the implications of protein structure in protein functional analysis and rational drug design as well as the limitations of existing experimental techniques to determine protein structure, computational approaches to predict protein structure will continue to be a necessity. The MULTICOM protein structure prediction pipeline stands ready to meet the needs of the research community and is accessible via a web service. The method uses a multi-level combination technique to combine multiple protein structure templates and sources of structural information to generate models and then employs a number of model refinement and selection tools to return the best possible predicted structure. The MULTICOM system is capable of using both template-based and template-free modeling to handle the full spectrum of protein modeling and generate predictions for all protein structure prediction tasks from the relatively easy to difficult. The system has been thoroughly and successfully tested in CASP8 and CASP9 and assessed as one of the best public, CASP-certified protein structure prediction servers.

6 Notes

1. An *e*-value is generated when using a search tool like BLAST [20, 21] to search the query against the template library. Usually, a low *e*-value means that the template has high similarity to the query.
2. Regions of a protein model usually refer to continuous segments of amino acids. Two regions have consistent structures if the similarity score between them is higher than a set threshold. The similarity score is calculated using the GDT-TS score generated from TM-score [17] when comparing them. In the MULTICOM system, we set the threshold to 0.75 for comparison of two regions.
3. Very short unaligned regions mean that there are less than ten residues unaligned in the template.
4. The absolute quality score of the model is the GDT-TS score between this model and its native structure. The GDT-TS score describes the expected similarity between the model and the native structure.
5. This approach is very sensitive about the input model pool. When the input model pool is small or contains many poor models, this approach does not work very well.
6. Two models are similar if the pairwise GDT-TS score is higher than a threshold. MULTICOM uses a threshold of 0.7 for comparison of two models.
7. TM-score [17], RMSD (average root mean square distance between the corresponding atoms), and GDT-TS score are commonly used tools to compare and evaluate protein structure predictions. The online version of the TM-score tool is available at http://zhanglab.ccmb.med.umich.edu/TM-score/. To compare the native structure (e.g., native.pdb) with a predicted structure (e.g., predicted.pdb), the predicted.pdb file is uploaded as Structure 1 and native.pdb is uploaded as Structure 2, leaving the e-mail address field blank. After running the comparison, the assessment results page shows the TM-score value.

Acknowledgment

The work was supported by an NIH grant R01GM093123 to J.C.

References

1. Anfinsen CB, Haber E, Sela M, White F Jr (1961) The kinetics of formation of native ribonuclease during oxidation of the reduced polypeptide chain. Proc Natl Acad Sci USA 47(9):1309
2. Chothia C, Lesk AM (1986) The relation between the divergence of sequence and structure in proteins. EMBO J 5(4):823
3. Bujnicki JM (2005) Protein-structure prediction by recombination of fragments. Chembiochem 7(1):19–27
4. Krieger E, Nabuurs SB, Vriend G (2003) Homology modeling. Methods Biochem Anal 44:509–524
5. Moult J, Pedersen JT, Judson R, Fidelis K (2004) A large-scale experiment to assess protein structure prediction methods. Proteins 23(3):ii–v
6. Cheng J (2008) A multi-template combination algorithm for protein comparative modeling. BMC Struct Biol 8(1):18
7. Fischer D (2003) 3D-SHOTGUN: a novel, cooperative, fold-recognition meta-predictor. Proteins Struct Funct Bioinf 51(3):434–441
8. Sali A, Blundell T (1994) Comparative protein modelling by satisfaction of spatial restraints. Proteins 64:C86
9. Wang Z, Eickholt J, Cheng J (2010) MULTICOM: a multi-level combination approach to protein structure prediction and its assessments in CASP8. Bioinformatics 26(7): 882–888
10. Cheng J, Wang Z, Tegge AN, Eickholt J (2009) Prediction of global and local quality of CASP8 models by MULTICOM series. Proteins 77(S9):181–184
11. Wang Z, Tegge AN, Cheng J (2008) Evaluating the absolute quality of a single protein model using structural features and support vector machines. Proteins 75(3):638–647
12. Moult J, Fidelis K, Kryshtafovych A, Rost B, Tramontano A (2009) Critical assessment of methods of protein structure prediction: Round VIII. Proteins 77(S9):1–4
13. Pettersen EF, Goddard TD, Huang CC, Couch GS, Greenblatt DM, Meng EC, Ferrin TE (2004) UCSF Chimera: a visualization system for exploratory research and analysis. J Comput Chem 25(13):1605–1612
14. LLC DS The PyMOL molecular graphics system. http://pymol.sourceforge.net/
15. Sayle R (1994) RasMol v2. 5-Molecular visualisation program
16. Jmol: an open-source Java viewer for chemical structures in 3D. http://jmol.sourceforge.net/. Accessed 10 Dec 2008
17. Zhang Y, Skolnick J (2004) Scoring function for automated assessment of protein structure template quality. Proteins 57(4):702–710
18. Li J, Deng X, Eickholt J, Cheng J (2013) Designing and benchmarking the MULTICOM protein structure prediction system. BMC Struct Biol 13:2
19. Berman H, Westbrook J, Feng Z, Gilliland G, Bhat T, Weissig H, Shindyalov I, Bourne P (2000) The protein data bank. Nucleic Acids Res 28(1):235–242
20. Altschul SF, Madden TL, Schäffer AA, Zhang J, Zhang Z, Miller W, Lipman DJ (1997) Gapped BLAST and PSI-BLAST: a new generation of protein database search programs. Nucleic Acids Res 25(17):3389–3402
21. Altschul S, Gish W, Miller W, Myers E, Lipman D (1990) Basic local alignment search tool. J mol Biol 215(3):403–410
22. Biegert A, Söding J (2009) Sequence context-specific profiles for homology searching. Proc Natl Acad Sci USA 106(10):3770–3775
23. Hughey R, Krogh A (1995) SAM: sequence alignment and modeling software system. In: Technical Report: UCSC-CRL-95-07. University of California at Santa Cruz
24. Finn RD, Clements J, Eddy SR (2011) HMMER web server: interactive sequence similarity searching. Nucleic Acids Res 39(Suppl 2):W29–W37
25. Soding J, Biegert A, Lupas A (2005) The HHpred interactive server for protein homology detection and structure prediction. Nucleic Acids Res 33(Web Server Issue):W244–W248
26. PRC, the profile comparer. http://supfam.org/PRC/
27. Sadreyev R, Grishin N (2003) COMPASS: a tool for comparison of multiple protein alignments with assessment of statistical significance. J Mol Biol 326(1):317–336
28. Zhang Y, Skolnick J (2005) TM-align: a protein structure alignment algorithm based on the TM-score. Nucleic Acids Res 33(7):2302–2309
29. Zhou H, Zhou Y (2005) SPEM: improving multiple sequence alignment with sequence profiles and predicted secondary structures. Bioinformatics 21(18):3615–3621
30. Fiser A, Sali A (2003) Modeller: generation and refinement of homology-based protein structure models. Methods Enzymol 374:461–491

31. Cheng J, Wang Z, Eickholt J, Deng X (2011) Recursive protein modeling: a divide and conquer strategy for protein structure prediction and its case study in CASP9. In: Bioinformatics and Biomedicine Workshops (BIBMW). IEEE. p 352–357
32. Leaver-Fay A, Tyka M, Lewis SM, Lange OF, Thompson J, Jacak R, Kaufman K, Renfrew PD, Smith CA, Sheffler W (2011) ROSETTA3: an object-oriented software suite for the simulation and design of macromolecules. Methods Enzymol 487:545–574
33. Wang Z, Tegge AN, Cheng J (2009) Evaluating the absolute quality of a single protein model using structural features and support vector machines. Proteins 75(3):638–647
34. Cheng J, Randall A, Sweredoski M, Baldi P (2005) SCRATCH: a protein structure and structural feature prediction server. Nucleic Acids Res 33(Web Server Issue):W72–W76
35. Tegge AN, Wang Z, Eickholt J, Cheng J (2009) NNcon: improved protein contact map prediction using 2D-recursive neural networks. Nucleic Acids Res 37(Suppl 2):W515–W518
36. Cheng J, Baldi P (2005) Three-stage prediction of protein β-sheets by neural networks, alignments and graph algorithms. Bioinformatics 21(Suppl 1):i75–i84

Chapter 4

Modeling of Protein Side-Chain Conformations with RASP

Zhichao Miao, Yang Cao, and Taijiao Jiang

Abstract

Modeling of side-chain conformations on a fixed protein backbone, also called side-chain packing, plays an important role in protein structure prediction, protein design, molecular docking, and functional analysis. RASP, or RApid Side-chain Predictor, is a recently developed program that can model protein side-chain conformations with both high accuracy and high speed. Moreover, it can generate structures with few atomic clashes. This chapter first provides a brief introduction to the principle and performances of the RASP package. Then details on how to use RASP programs to predict protein side-chain conformations are elaborated. Finally, it describes case studies for structure refinement in homology modeling and residue substitution.

Key words Protein structure prediction, Side-chain packing, Rotamer library, Energy function, Combinatorial search, RASP, CIS-RR

1 Introduction

A protein adopts specific side-chain conformations while folding into a particular structure and carrying out its function. Thus, accurate and rapid modeling of protein side-chain conformations is a crucial step in protein design [1] and protein structure modeling as well as protein function and interaction analysis. During the last two decades, many efforts have been dedicated to the prediction of protein side-chain conformations [2–17].

Protein side-chain conformations that are mainly determined by dihedral torsion angles are called rotational isomers or rotamers [18, 19]. As certain rotamer dihedral angles are statistically more favorable, they can be clustered into discrete conformations (e.g., t, g+, or g− conformations). Therefore, in order to significantly reduce computational expense, side-chain conformation prediction problems can generally be reduced to a combinatorial search problem based on discrete rotamers. To carry out an efficient search through the enormous number of rotamer combinations, enormous search algorithms guided by energy/scoring functions

Daisuke Kihara (ed.), *Protein Structure Prediction*, Methods in Molecular Biology, vol. 1137,
DOI 10.1007/978-1-4939-0366-5_4, © Springer Science+Business Media New York 2014

have been published including (1) dead-end elimination [20], (2) simulated annealing [21], (3) Monte Carlo [11], (4) A* [22], (5) integer programming [23], (6) self-consistent mean field [24], and (7) graph theory-based approaches [2, 25]. Combination of these search algorithms is critical in side-chain prediction. For example, to achieve high-speed SCWRL4 [9] and SCATD [17] combine DEE, branch-and-bound, and tree decomposition searches. The recently developed energy scoring functions range from simple van der Waals potentials [4, 26] to more complicated ones incorporating hydrogen bonding [9], solvation [27], and statistical orientation terms [12].

As a result of previous efforts, the prediction of side-chain conformations has become more and more accurate. A further improvement of accuracy stems also from increased computational efficiency. For example, although the recently developed side-chain packing program SCRWL4 (Dunbrack's lab) and the program CIS-RR (short for clash-detection guided iterative search with rotamer relaxation, in our lab) display accuracy improvements of ~3 % in side-chain χ dihedral accuracy over SCRWL3, they are six times slower than the former SCRWL3 version, indicating the challenge in achieving both high accuracy and high speed in the prediction of protein side-chain conformations. Recently, we developed a RApid Side-chain Predictor, called RASP [28]. RASP combines two steps in modeling of protein side-chain conformations: (1) rapid generation of high-quality initial structures and (2) rapid elimination of atomic clashes.

RASP has been compared with some well-established programs, including SCWRL4 [9], OPUS-Rota [29], and IRECS [6], by using the SCWRL4 test set. As shown in Table 1, for prediction

Table 1
Comparison of RASP and CIS-RR with some recently developed side-chain prediction programs by using the SCWRL4 test set

Program	Time (min)	Clashes	χ_1 (%)	χ_{1+2} (%)	RMSD (Å)
RASP	1.8	47	85.1	74.7	1.5
CIS-RR	73	59	84.9	74.9	1.5
SCWRL4	33	411	85.0	75.4	1.5
SCWRL3	5	1,107	82.2	71.3	1.6
OPUS-Rota	26	623	85.0	75.0	1.4
IRECS	38	1,201	83.6	71.8	1.7

Correctness percentage of χ_1 is defined as the percentage of residues whose predicted χ_1 dihedral is within 40° of the χ_1 dihedral of native side chains, while correctness percentage of χ_{1+2} is defined as the percentage of residues for which both χ_1 and χ_2 are within 40° of those of native side chains accuracy in terms of correctness percentages for χ_1 and χ_{1+2} dihedrals and side-chain root mean square deviation (RMSD), RASP and CIS-RR are comparable to these side-chain prediction programs. For speed, RASP is 14 times faster than OPUS-Rota and 18 times faster than SCWRL4. Moreover, it generates fewer clashes than SCWRL4 and OPUS-Rota.

2 Materials

2.1 Input Data

The Brookhaven PDB [30] format is used for input protein files. Main-chain structure data should be included in the input file, especially for the three main-chain atoms (N, CA, and C). But if the main-chain oxygen (O) atom is missing, RASP will generate its coordinates according to the peptide plane using default parameters. Some uncommon residue types are assimilated to the standard amino acid type that is most similar to it. For example, the residue type MSE (selenomethionine) is deemed as methionine in the prediction, while SMC (S-methylcysteine) is assimilated to cystine.

2.2 Programs in the RASP Package

The RASP package includes RASP, CIS-RR, and RASP tools.

RASP is a super fast and highly accurate side-chain packing program that is suitable for high-throughput structure modeling and optimization.

CIS-RR is a user-friendly program for accurate side-chain prediction that is suitable for protein residue substitution modeling.

RASP tools are used for measuring side-chain structural information and include three tools:

- RASPsym flips symmetric residues in the PDB to the same side. For instance, the OD1 and OD2 atoms in aspartic acid are symmetric (Fig. 1):

  If the names of the two atoms are exchanged, the side-chain conformation does not change. However, since only one atom (OD1) determines its dihedral angle, the structure needs to be flipped before measuring the dihedral angle. Otherwise, the measured angle would be 180° away.

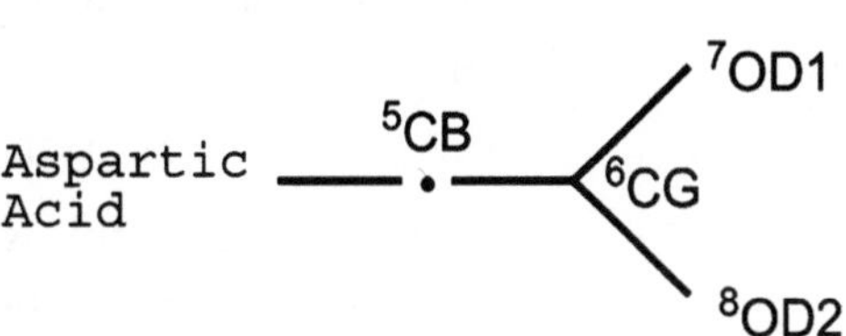

Fig. 1 Symmetric side chain

- RASPchi measures the dihedral angles of all the residues in a PDB structure.
- RASPrms calculates the RMSD between two PDB structures.

2.3 Website

The RASP package (including the examples) is available at http://jianglab.ibp.ac.cn/lims/rasp/rasp.

3 Methods

Please note that programs in the RASP package run on Linux systems with the gcc library.

3.1 RASP Programs

3.1.1 Platform

Several Linux platforms have been tested, including

- gcc version 4.4.5 20100728 (prerelease) (Debian 4.4.4-8).
- gcc version 4.4.3 (Ubuntu 4.4.3-4ubuntu5).
- gcc version 4.3.2 (Debian 4.3.2-1.1).
- gcc version 4.1.2 20061115 (prerelease) (Debian 4.1.1-21).
- gcc version 4.6.3 (Ubuntu/Linaro 4.6.3-1ubuntu5).
- gcc on Linux Mint 64.

3.1.2 Installation

Uncompress the package to the directory where you want to install the program, and then double click on the "setup" file. If the setup does not run, make sure that it can be executed by using the command "chmod+x setup_rasp." After installation, four files appear in the directory:

1. bbdep11.bin: the side-chain rotamer library file.
2. RASP.ini: the parameter configuration file that includes useful parameters.
3. README: a brief introduction to RASP.
4. RASP: the executive binary file for side-chain structure prediction.

After installation, "bbdep11.bin" and "RASP.ini" should remain in the directory, while the executive file RASP can be copied to any directory.

3.1.3 Usage

1. For side-chain modeling based on main-chain structure and sequence:

```
./RASP -i [input_pdb] -o [output_pdb]
```

The "input_pdb" is the input main-chain structure file, based on which RASP performs the prediction. "output_pdb" is any name assigned for the output file (*see* **Notes 1** and **2**). As an example, "-o abcd.pdb" generates a PDB file named "abcd.pdb" as a result.

2. Modeling side-chain conformations for mutated residues: This function was designed for the modeling of mutated sequences. A sequence file is provided with the "-s" option to specify mutated and fixed residues. Once a sequence is given, RASP predicts side-chain conformations according to the given sequence other than the sequence contained in the input main-chain PDB file (*see* **Notes 3** and **4**). In the sequence, lower case letters indicate residues to be fixed with their original conformations, while upper case letters need to be predicted.

```
./RASP -i 1A8Q.pdb -o result.ent -s substitution.fasta
```

```
An example of substitution.fasta:
>1A8Q:A|PDBID|CHAIN|SEQUENCE
PICTTRDGVEIFYKDWGQGRPVVFIHGWPLNGdawqdQLKAVVDAGYRGIAHDRRGHGHSTP
VWDGYDFDTFADDLNDLLTDLDLRDVTLVAHSMGGGELARYVGRHGTGRLRSAVLLSAIPPV
MIKSDKNPDGVPDEVFDALKNGVLTERSQFWKDTAEGFFSANRPGNKVTQGNKDAFWYMAM
AQTIEGGVRCVDAFGYTDFTEDLKKFDIPTLVVHGDDDQVVPIDATGRKSAQIIPNAELKVYEGS
SHGIAMVPGDKEKFNRDLLEFLNK
```

In this case, the "**dawqd**" region will take their side-chain conformations as in the input file.

3. Using ligand coordinates for spatial constraints: If the coordinates of the protein's ligands or water molecules around the protein are provided as spatial constraints, use "-f" for the ligand coordinates file (*see* **Note 5**). As the ligands can be any kind of molecules, they are read in atom by atom. The ligand coordinates file should be in PDB format headed by "ATOM" or "HETATM" in each line. For instance:

```
./RASP -i abcd.pdb -o efgh.pdb -f ijkl.pdb
```

RASP will predict side-chain structures on the abcd.pdb main-chain structure using ijkl.pdb as spatial constraints and preserve the result structure in efgh.pdb.

4. Recording predicted side-chain conformations: RASP also has a "-d" option to record all the dihedral angles in a ".dihed" file with the following format (Fig. 2):

As a case:

```
./RASP -i abcd.pdb -o efgh.pdb -d
```

The predicted side-chain dihedral angles will be stored in "efgh.pdb.dihed."

```
DIHEDRAL ANGLES
-------------------------------------------------------------
Res|C|   No|        Phi|     Psi|    Chi1|    Chi2|    Chi3|    Chi4|
-------------------------------------------------------------
SER B     9     999.00  -44.39   177.67    0.00     0.00     0.00
ILE B    10     -52.04  -30.30  -171.53  179.39     0.00     0.00
ASN B    11     -52.23  -34.37   -45.73   46.12     0.00     0.00
GLN B    12     -71.58  -42.12   160.31   72.43   -93.24     0.00
LYS B    13     -58.21  -43.15   -94.95  149.88   173.72   161.43
LEU B    14     -58.22  -58.78   -99.52  118.87     0.00     0.00
ALA B    15     -46.06  -47.24        -       -        -        -
LEU B    16     -55.13  -53.80  -170.01  -48.45     0.00     0.00
VAL B    17     -53.58  -29.61   178.84    0.00     0.00     0.00
ILE B    18     -75.92  -27.19   -70.65  175.23     0.00     0.00
LYS B    19     -86.74  -39.69    76.80  151.95  -176.67    64.18
SER B    20     -92.52  -61.14    38.25    0.00     0.00     0.00
GLY B    21     -95.27  174.69        -       -        -        -
LYS B    22     -91.63  153.63   -64.79  112.74    76.27   175.48
TYR B    23    -149.57 -174.70    67.35  -80.37     0.00     0.00
THR B    24    -160.70  156.13   157.80    0.00     0.00     0.00
```

Fig. 2 Dihedral angle file format

3.2 CIS-RR Program

3.2.1 Website

The CISRR package is available at http://jianglab.ibp.ac.cn/lims/cisrr/cisrr.html.

3.2.2 Installation

Extract the CISRR.tar.gz, and place the CISRR folder anywhere without modifying the newly built subdirectories and files. To use the software in a different directory, specify the program path in the command: <program path>/bin/CISRR. CIS-RR is mainly used for residue substitution analysis.

3.2.3 Usage

`./CISRR -i [input_pdb] -o [output_pdb] -m [mutation information].`

The "-m" indicates an amino acid substitution, formatted as [chain id] [sequence number] [original residue name (one letter)] [new residue name (one letter)]. If more than two mutations are needed, use more of the "-m" options one by one.

For example, "CISRR -i 1AGI.pdb -o 1AGI_sp.pdb -m A 5 Y W" is used to make a substitution from TYR to TRP at site 5 of chain A.

3.3 RASP Tools

1. `./RASPsym [input_pdb] [output_pdb]`

   `Example: ./RASPsym abcd.pdb efgh.pdb`

   Asp, Glu, Phe, Tyr, and Arg will be flipped to the same side (in efgh.pdb).

2. `./RASPchi [input_pdb]`

   `Example: ./RASPchi 1T0K.pdb`

The side-chain dihedrals will be measured and stored in 1T0K.pdb.dihed.

To use RASPchi, just put the PDB file name after RASPchi in the command line. A file with suffix ".dihed" will be generated. This file adopts the same format as the ".dihed" file generated by the RASP "-d" option (format shown in Fig. 1). Data in this file can be used to compare side-chain conformations.

3. `./RASPrms [PDB1] [PDB2] (options)`

   To compare the RMSD of side-chain coordinates, RASPrms can be used. The options in RASPrms could be "-ca," "-mc," "-ss," "-sf," and "-tot." The "-ca" option calculates only the RMSD for CA atoms; "-mc" measures the RMSD of the four main-chain atoms (N, CA, C, and O); "-ss" measures the side-chain RMSD excluding the main-chain atoms (main chain are deemed as the same); "-sf" option calculates RMSD for the whole structure without superimposition; "-tot" superimposes all of the coordinates and calculates the total RMSD.

As a case:

```
./RASPrms 1E7K.pdb 1T0K.pdb -ca -ss -tot
```

RMSD for CA atoms, for side-chain atoms, and for the whole structure are measured. The results will be printed on the screen. RASPrms has only simple functionalities. For more complex ones, please refer to Profit (http://www.bioinf.org.uk/profit/).

3.4 Case Studies

3.4.1 Structure Refinement in Homology Modeling

To use the modeled structure for in-depth protein function analysis such as binding to ligands or other proteins, a high-quality prediction of side-chain conformations is required. However, for highest speed, homology modeling programs usually use very simple rotamer models and scoring functions that might generate excessive atomic clashes through unfavorable side-chain conformations. RASP offers very convenient and useful tools to optimize modeled structures.

Here, we provide a real case for protein structure modeling to illustrate the application of RASP. L30e in Eukaryotes (1T0K) [31] and 15.5 kDa (1E7K) [32] in humans are homologous RNA-binding proteins. Sequence identity between the two proteins is 16.7 %, and the RMSD for Cα atoms is 1.76. Therefore, 15.5 kDa protein could be used as a homologous template for modeling the L30e protein (Fig. 3).

First, sequences of the two proteins need to be aligned. If no insertions or deletions (gaps) are in the alignment, we can directly use RASP to model the structure. As in this case, Modeller [33] can be employed to generate a structural model (name it model_1) with insertions and deletions. L30e protein binds to ribosomal

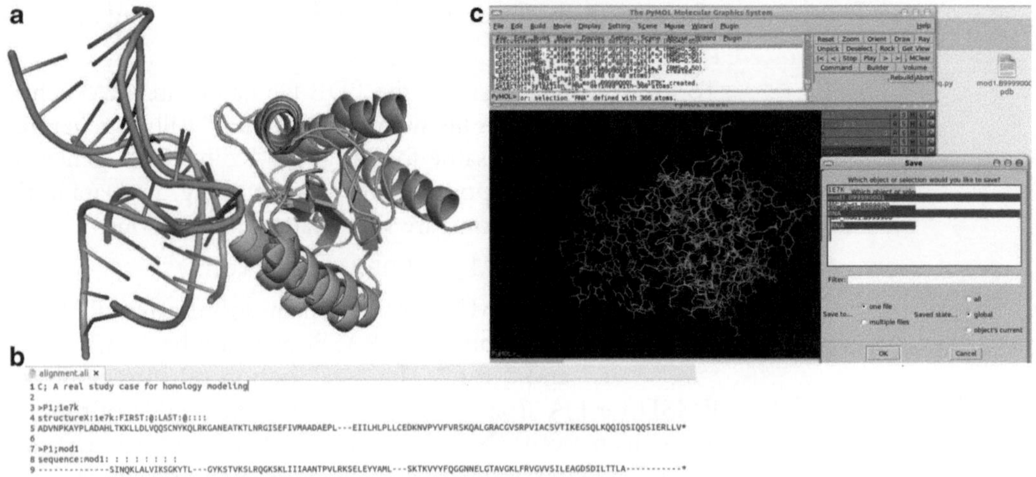

Fig. 3 Steps in structure modeling. (**a**) Superimposition between L30e and 15.5 kDa protein. (**b**) Sequence alignment used for homology modeling. (**c**) Superimpose the structure generated by Modeller to the RNA–protein complex, and save the RNA coordinates as spatial constraints

RNA, and 15.5 kDa protein binds to U4 kink-turn; both of the protein-binding regions adopt kink-turn structural module. Thus, the known RNA coordinates could be used as spatial constraints to optimize the side-chain conformations. We superimpose model_1 to the template 15.5 kDa protein (saved as model_2) and preserve the U4 snoRNA coordinates (saved as RNA.pdb). Finally, we use RASP to rebuild side-chain conformations on model_1:

```
./RASP -i model_2 -o L30e.model.pdb -s L30e.seq -f RNA.pdb.
```

Then, the side-chain conformations can be assessed using RASPsym and RASPchi. Compared with the native structure (1TOK), the side-chain conformation modeled by Modeller is 28.4 % for χ_1 dihedral and 53.1 % for χ_2. After optimization by RASP the accuracy improves to 35.8 and 58.0 %, respectively. Besides, this improvement only takes ~0.2 s. Therefore, we conclude that RASP can greatly improve the accuracy of protein homology structures.

3.4.2 Residue Substitution Analysis

Protein residue substitution has little impact on the main-chain conformation in most cases [34]. It indicates that protein residue substitutions can be modeled by side-chain packing. CIS-RR is an accurate and user-friendly program for residue substitution modeling. In an unpublished study of discovering drug resistance mutations of influenza neuraminidase (NA), we employed CIS-RR to run a high-throughput modeling of 19 types of residue substitution at the sites of drug (oseltamivir)-binding regions:

```
For residue R at site i of chain A
For each amino acid type N
./CISRR -i 3CL2.pdb -o 3CL2_RiN.pdb -m A i R N
End-for N
End-for i
```

Then we calculated the binding free energy changes. Drug resistance mutations are regarded to be those whose energies decrease in binding with oseltamivir and increase or stabilize in binding with substrate. The prediction covers many of the reported drug resistance mutations, such as H274Y/F [35, 36], E119A/G/V [37–40], I222V/M/K/R [39–43], and D151N/G [44]. More importantly, we discovered some novel drug resistance mutations that were validated by experiments. This case suggests that CIS-RR is a useful program for high-throughput residue substitution modeling.

4 Notes

1. Can the input PDB file and the output PDB file be of the same name?

   If they use the same name, the input file will be overwritten by the output file. Caution: If you do not want to eliminate the input PDB file, do not use the input PDB file name for the output PDB file.
2. How about the atom order in a residue of the output of RASP?

   The output order of the atoms in a residue follows the standard PDB format (http://www.rcsb.org/pdb/file_formats/pdb/pdbguide2.2/PDB_format_1992.pdf).
3. When a letter in the given sequence file (–s) is in lower case but the residue type does not match the residue type in the input PDB, what will RASP do?

   This residue is regarded as a mutation, which will also be predicted according to the residue type given in the sequence file.
4. When a letter in the given sequence file (–s) is in lower case but the residue in the PDB file is incomplete (some atoms are missing), what will RASP do?

   RASP detects such missing atoms and predicts this residue with a new conformation.
5. Can "-s," "-f," and "-d" be used together?

   Yes, they can be used in any combination.

Acknowledgments

We gratefully thank Pascal Auffinger from Institut de Biologie Moléculaire et Cellulaire, Centre national de la recherche scientifique, for his help on critical editing of the manuscript.

References

1. Canzar S, Toussaint NC, Klau GW (2011) An exact algorithm for side-chain placement in protein design. Optimization Letters 5(3):393–406
2. Canutescu A, Shelenkov A, Dunbrack R (2003) A graph-theory algorithm for rapid protein side-chain prediction. Protein Sci 12(9):2001–2014
3. Cao Y et al (2011) Improved side-chain modeling by coupling clash-detection guided iterative search with rotamer relaxation. Bioinformatics 27(6):785–790
4. Bower M (1997) Prediction of protein side-chain rotamers from a backbone-dependent rotamer library: a new homology modeling tool. J Mol Biol 267(5):1268–1282
5. Fromer M et al (2010) SPRINT: side-chain prediction inference toolbox for multistate protein design. Bioinformatics (Oxford, England) 26(19):2466–2467
6. Hartmann C, Antes I, Lengauer T (2007) IRECS: a new algorithm for the selection of most probable ensembles of side-chain conformations in protein models. Protein Sci 16(7):1294–1307
7. Wang C, Schueler-Furman O, Baker D (2005) Improved side-chain modeling for protein-protein docking. Protein Sci 14(5):1328–1339
8. Petrella R, Karplus M (2001) The energetics of off-rotamer protein side-chain conformations. J Mol Biol 312(5):1161–1175
9. Krivov G, Shapovalov M, Dunbrack R (2009) Improved prediction of protein side-chain conformations with SCWRL4. Proteins 77(4):778–795
10. Liang S, Grishin NV (2002) Side-chain modeling with an optimized scoring function. Protein Sci 11(2):322–331
11. Liang S et al (2011) Protein side chain modeling with orientation-dependent atomic force fields derived by series expansions. J Comput Chem 32(8):1680–1686
12. Lu M, Dousis A, Ma J (2008) OPUS-PSP: an orientation-dependent statistical all-atom potential derived from side-chain packing. J Mol Biol 376(1):288–301
13. McGregor M, Islam S, Sternberg M (1987) Analysis of the relationship between side-chain conformation and secondary structure in globular proteins. J Mol Biol 198(2):295–310
14. Peterson RW, Dutton PL, Wand AJ (2004) Improved side-chain prediction accuracy using an ab initio potential energy function and a very large rotamer library. Protein Sci 13(3):735–751
15. Tuffery P et al (1991) A new approach to the rapid determination of protein side chain conformations. J Biomol Struct Dyn 8(6):1267–1289
16. Xiang Z, Honig B (2001) Extending the accuracy limits of prediction for side-chain conformations1. J Mol Biol 311(2):421–430
17. Xu J (2005) Rapid protein side-chain packing via tree decomposition. In: Miyano S et al (eds) Research in computational molecular biology. Springer, Berlin, Heidelberg, pp 423–439
18. Dunbrack RL Jr, Cohen FE (1997) Bayesian statistical analysis of protein side-chain rotamer preferences. Protein Sci 6(8):1661–1681
19. Lovell SC et al (2000) The penultimate rotamer library. Proteins 40(3):389–408
20. Desmet J et al (1992) The dead-end elimination theorem and its use in protein side-chain positioning. Nature 356(6369):539–542
21. Lee C, Subbiah S (1991) Prediction of protein side-chain conformation by packing optimization. J Mol Biol 217(2):373–388
22. Leach AR, Lemon AP (1998) Exploring the conformational space of protein side chains using dead-end elimination and the A* algorithm. Proteins 33(2):227–239
23. Kingsford L, Bernard C, Mona S (2005) Solving and analyzing side-chain positioning problems using linear and integer programming. Bioinformatics 21(7):1028–1039
24. Lee C (1994) Predicting protein mutant energetics by self-consistent ensemble optimization. J Mol Biol 236(3):918–939
25. Samudrala R, Moult J (1998) An all-atom distance-dependent conditional probability discriminatory function for protein structure prediction. J Mol Biol 275(5):895–916

26. Vásquez M (1995) An evaluation of discrete and continuum search techniques for conformational analysis of side chains in proteins. Biopolymers 36(1):53–70
27. Mendes J et al (2001) Implicit solvation in the self-consistent mean field theory method: side chain modelling and prediction of folding free energies of protein mutants. J Comput Aided Mol Des 15(8):721–740
28. Miao Z, Cao Y, Jiang T (2011) RASP: rapid modeling of protein side chain conformations. Bioinformatics 27(22):3117–3122
29. Lu M, Dousis A, Ma J (2008) OPUS-Rota: a fast and accurate method for side-chain modeling. Protein Sci 17(9):1576–1585
30. Berman HM et al (2000) The protein data bank. Nucleic Acids Res 28(1):235–242
31. Chao J, Williamson J (2004) Joint X-Ray and NMR refinement of the yeast L30e-mRNA complex. Structure 12(7):1165–1176
32. Vidovic I et al (2000) Crystal structure of the spliceosomal 15.5 kDa protein bound to a U4 snRNA fragment. Molecular cell 6(6): 1331–1342
33. Eswar N et al (2006) Comparative protein structure modeling using modeller, in current protocols in bioinformatics. John Wiley & Sons, Inc. http://onlinelibrary.wiley.com/doi/10.1002/0471250953.bi0506s15/abstract
34. Chothia C, Lesk AM (1986) The relation between the divergence of sequence and structure in proteins. EMBO J 5(4):823–826
35. Le QM et al (2005) Avian flu: isolation of drug-resistant H5N1 virus. Nature 437(7062):1108
36. Wang MZ, Tai CY, Mendel DB (2002) Mechanism by which mutations at his274 alter sensitivity of influenza a virus n1 neuraminidase to oseltamivir carboxylate and zanamivir. Antimicrob Agents Chemother 46(12): 3809–3816
37. Ilyushina NA et al (2010) Effect of neuraminidase inhibitor-resistant mutations on pathogenicity of clade 2.2 A/Turkey/15/06 (H5N1) influenza virus in ferrets. PLoS Pathog 6(5):e1000933
38. Abed Y, Goyette N, Boivin G (2004) A reverse genetics study of resistance to neuraminidase inhibitors in an influenza A/H1N1 virus. Antivir Ther 9(4):577–581
39. Deng YM et al (2011) A comparison of pyrosequencing and neuraminidase inhibition assays for the detection of oseltamivir-resistant pandemic influenza A(H1N1) 2009 viruses. Antiviral Res 90(1):87–91
40. Pizzorno A et al (2011) Generation and characterization of recombinant pandemic influenza A(H1N1) viruses resistant to neuraminidase inhibitors. J Infect Dis 203(1):25–31
41. van der Vries E, Stelma FF, Boucher CA (2010) Emergence of a multidrug-resistant pandemic influenza A (H1N1) virus. N Engl J Med 363(14):1381–1382
42. Boltz DA et al (2010) Emergence of H5N1 avian influenza viruses with reduced sensitivity to neuraminidase inhibitors and novel reassortants in Lao People's Democratic Republic. J Gen Virol 91(Pt 4):949–959
43. Nguyen HT et al (2010) Recovery of a multidrug-resistant strain of pandemic influenza A 2009 (H1N1) virus carrying a dual H275Y/I223R mutation from a child after prolonged treatment with oseltamivir. Clin Infect Dis 51(8):983–984
44. Okomo-Adhiambo M et al (2010) Host cell selection of influenza neuraminidase variants: implications for drug resistance monitoring in A(H1N1) viruses. Antiviral Res 85(2):381–388

Chapter 5

Direct Coupling Analysis for Protein Contact Prediction

Faruck Morcos, Terence Hwa, José N. Onuchic, and Martin Weigt

Abstract

During evolution, structure, and function of proteins are remarkably conserved, whereas amino-acid sequences vary strongly between homologous proteins. Structural conservation constrains sequence variability and forces different residues to coevolve, i.e., to show correlated patterns of amino-acid occurrences. However, residue correlation may result from direct coupling, e.g., by a contact in the folded protein, or be induced indirectly via intermediate residues. To use empirically observed correlations for predicting residue–residue contacts, direct and indirect effects have to be disentangled. Here we present mechanistic details on how to achieve this using a methodology called *Direct Coupling Analysis* (DCA). DCA has been shown to produce highly accurate estimates of amino-acid pairs that have direct reciprocal constraints in evolution. Specifically, we provide instructions and protocols on how to use the algorithmic implementations of DCA starting from data extraction to predicted-contact visualization in contact maps or representative protein structures.

Key words Direct coupling analysis, Maximum entropy, Contact prediction, Residue–residue interactions, Coevolution, Direct correlations, Statistical inference

1 Introduction

In the course of evolution, structure, and function of proteins are remarkably conserved, whereas amino-acid sequences vary strongly between homologous, i.e., evolutionary-related proteins. However, structural conservation constrains this sequence variability and forces different residues to *coevolve*: Acceptable amino-acid substitutions in one site depend on the amino-acid composition of other sites that are spatially close in the three-dimensional protein structure (even if possibly distant along the sequence). Neighboring residues tend to vary in a correlated way.

Recent advances in genomic sequencing have provided enough data to perform evolutionary analyses at the level of residue covariation in single protein families. It is therefore an important challenge in computational biology to exploit computationally detected covariation for structural prediction of proteins. However, the task is nontrivial: One of the major obstacles in understanding residue

Daisuke Kihara (ed.), *Protein Structure Prediction*, Methods in Molecular Biology, vol. 1137,
DOI 10.1007/978-1-4939-0366-5_5, © Springer Science+Business Media New York 2014

covariation is to separate the amalgam of signals that results from different types of correlations. This is further complicated by the quality of the data, which can be incomplete (small number of sequences) and unbalanced (e.g., high redundancy), and by the effects of phylogeny. Several research efforts dealt with residue correlations with mixed success [1–6]. However, recent methodologies mainly based on global statistical models aim to disentangle different sources of correlations in order to obtain what is called *direct statistical couplings between residue positions*, which turn out to be much more accurate predictors of residue physical contacts [7–13] than sheer correlation measures. The basic idea is very intuitive: Empirical correlations between positions in a protein family, more precisely in a multiple-sequence alignment (MSA) describing this family, may well be induced by a direct coupling, but they may also result from indirect couplings via one or more intermediate residues.

Here we present the mechanistic details on how to infer direct residue-pair couplings using a methodology called *Direct Coupling Analysis* (DCA). DCA has been shown to produce highly accurate estimates of amino-acid pairs that are directly coupled through reciprocal constraints in evolution [7, 9]. Specifically, we provide instructions and protocols on how to use the algorithmic implementations of DCA starting from data extraction to contact visualization. Such software implementations have been made available to the public and the main purpose of this chapter is to provide guidance and facilitate its use to the scientific community. For a mathematical derivation of the *mean-field* formulation of DCA, utilized here, please refer to [7].

DCA is a powerful statistical tool that has been useful to study structural features of single domain proteins, the organization of oligomers, conformational variance of proteins, detailed features of protein–protein interactions [7, 14–16], as well as de novo protein structure prediction [17–20]. Its full potential can only be exploited if this tool is available and easy to use for a larger number of scientists.

2 Materials

In this section, we describe the minimum and optional input data sets as well as the software tools required for protein contact prediction using mean-field DCA.

2.1 Input Data

2.1.1 Multiple-Sequence Alignments (MSA)

The most important input data set is an MSA of proteins belonging to a given family. Here we focus on alignments collected using Profile Hidden Markov Models (HMM) in the Pfam domain database [21]. Such alignments are being updated continuously and are freely and easily accessible to the scientific community via http://pfam.sanger.ac.uk/ or any of the other Pfam mirrors. The software tools presented here use fasta format for the MSA.

2.1.2 Protein Structure 3D Coordinates

This is an optional data set used to validate contact predictions. If a given protein with 3D coordinates in the Protein Data Bank (PDB) [22] can be mapped to one or more of the Pfam domains, then we can verify or analyze the predicted residue–residue pairs directly in the 3D model of the protein. Protein coordinates can be accessed and downloaded using their corresponding PDB IDs at http://www.pdb.org.

2.2 Software Tools and Algorithmic Implementations

The core of our residue contact estimation methodology is the algorithmic implementation of mean-field DCA [7]. This implementation allows us to process more families with longer amino-acid chains and a larger number of sequences than the message passing formulation previously given in ref. [9]. *See* Subheading 4 for details on the requirements on input MSA. The mathematical derivation and a large-scale study showing its capabilities can be found in ref. [7]. The algorithm implementation is written in Matlab scripting language. The main reasons for choosing a scripting language like Matlab are its simplicity, the number of available bioinformatics tools, and its optimization for linear algebra computations. Faster implementations are possible using other non-scripting languages like C++; however, our current Matlab implementation is easier to read and its speed is acceptable for the vast majority of domain families.

2.2.1 Algorithm 1: mfDCA

Mean-field DCA is implemented in a Matlab script called `dca.m` which has two input parameters: (1) the name of the MSA input file (which has to be provided in fasta format) and (2) the name of the output file; internally there are two more parameters that could be modified: (1) the parameter theta which is a threshold on the value of sequence identity we use to define if sequences are considered independent or not and (2) the pseudo-count weight, which is a regularization term to prevent singularities due to insufficient sampling of rare amino-acid combinations.

2.2.2 Mapping HMM to Proteins

The output of Algorithm 1 will provide a list of residue–residue pairs that we can sort using its associated Direct Information (DI) metric. High values of DI tend to be representative of strong direct coupling and serve as predictors for residue–residue contacts. However, these pairs represent residues in the domain family associated with a given HMM and not connected to a specific protein of interest. More specifically, HMM matches usually do not start in the first amino acid of a protein and insert gaps where needed, so residue numbering in the alignment and in each protein is not the same. In order to predict specific contacts in a given protein, we need to map the HMM residues to a particular amino-acid sequence. If there is an experimental structure stored in the PDB, then usually a mapping has been made between an HMM and the protein. This mapping is depicted in the PDB site. After searching for an individual PDB ID, just go to *sequence tab* → *Annotations* → *Add annotation* → *Pfam*. It is also possible to do this mapping only with sequence information. To do this, we can turn to the HMMER software tools [23] for which we can use the tool `hmmscan` along with the protein sequence in fasta format and a local copy of the Pfam database or the HMM of a specific family. We will provide an example in Subheading 3.

2.2.3 Tool 1: Contact Map Visualization

Once we have a mapping between the DI ranked domain contacts predicted using mfDCA and a particular protein, we can visualize DCA contact maps using the script `plotDCAmap.m`. It has five input parameters: (1) a two column matrix with a given number of ranked DCA pairs, (2) an optional two column matrix with native protein contacts for comparison (shown in the upper triangular part of the map), (3) a vector with the protein residue range, (4) a flag to color the map by DCA ranking, and (5) a flag to plot a mirror image of the DCA map on the upper triangular map. The output is a Matlab figure showing the contact map in a grid according to the residue number and, if selected, contacts are colored based on a DCA ranking colormap. The figure can be exported to all Matlab supported image files. A sparse matrix with the non-zero contact values is the output from this script.

2.2.4 Tool 2: Contact Visualization in 3D Structure

We have also developed an optional simple script to visualize the predicted contacts directly on a 3D model of a protein. This requires the molecular modeling and visualization software Chimera [24] which is freely accessible for noncommercial purposes. `GeneratePseudobonds.bash` is a bash Unix shell script that uses a dependency written in AWK to produce files to be read by Chimera. The input of `GeneratePseudobonds.bash` is a file containing the protein residue pairs, the chain ID where we want to display contacts, and optionally an offset when using C-alpha models that have a different residue indexing. The output are two files: (1) a text file with the pseudo-bonds to be read in the pseudobond reader panel in Chimera and (2) a simple Chimera script to display such bonds.

2.3 Web Sites

There are three Web sites that contain information concerning DCA contact prediction:

1. http://dca.ucsd.edu and its mirror sites http://dca.rice.edu and http://dca.upmc.fr. These Web sites contain general information about DCA for contact prediction as well as for protein structure prediction, including reference to research articles involving DCA and access to some of the scripts described in this chapter. These sites also display news and updates about tools relevant to DCA.

2.4 Software Dependencies

Several of the tools and algorithms described earlier require specific software packages in order to be run properly. Here we list such packages that are a prerequisite of our methodology:

1. Matlab—an interactive environment for technical computing. It is required by the implementation of mfDCA (`dca.m`). The bioinformatics toolbox is also a requirement for this implementation.
2. HMMER—biosequence analysis using profile HMM. This is required to map HMM to protein sequences with or without an experimental structure from the PDB.
3. Bash/AWK—a standard Unix shell and an interpreted language required for simple scripting and data formatting. Other similar shell environments or languages could easily replace the needs for Bash in our methodology.
4. Chimera—a molecular graphics program for protein visualization developed by UCSF. It is needed for the visualization of contacts as links overlaid in the 3D representation of a particular protein structure.

3 Methods

The process of residue contact inference in domain families using DCA is summarized in Fig. 1. The process is organized into three stages: (1) inference, (2) validation, and (3) visualization. This section describes in detail each of these stages as well as the software tools required to perform the contact prediction task.

3.1 Inference

3.1.1 Data Retrieval

The first step in the inference task consists of obtaining the input data required for the alignment. As described in Subheading 3, the input data consists of an MSA of a given domain family. This will be the sole input dataset required by our software implementation of mean-field DCA. An MSA for specific protein families classified by Pfam can be downloaded directly from their site (http://pfam.sanger.ac.uk/) or via ftp. In this site we can search for a family by writing the accession or ID on the "JUMP TO" search box. Once we are in the family summary page, we retrieve this by selecting "alignments" and downloading the files under "Formatting options." There, select the format as "Fasta", sequence as "Inserts lower case," and "Gaps" as a mixture of "." and "-" characters. Once the MSA has been retrieved, it can be directly analyzed by `dca.m`. It is also possible to perform some preprocessing of the input data. For example, we may want to eliminate sequences with many continuous gaps "-" or starting or trailing gaps or to eliminate exact repeat sequences. We might also keep only sequences for

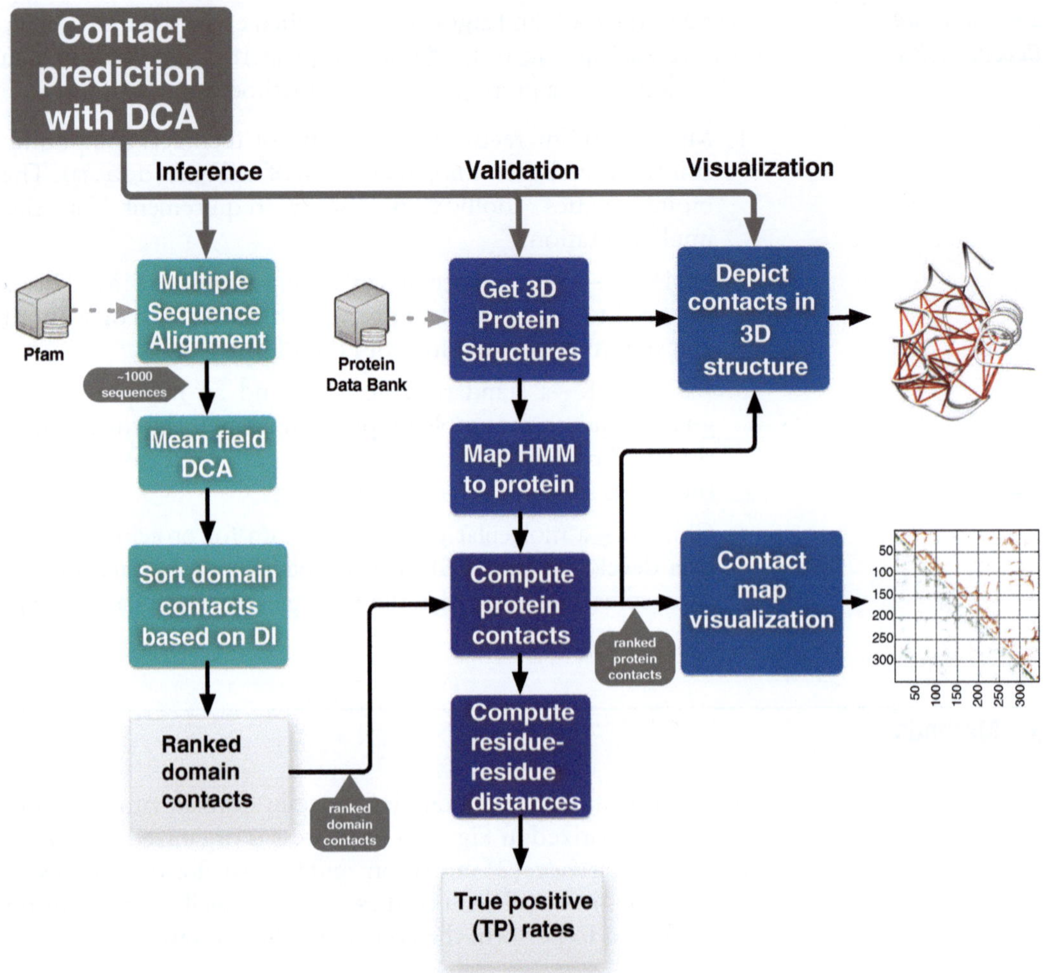

Fig. 1 The process of contact prediction is divided into three stages: (1) inference, where the MSA is processed by mfDCA in order to produce directly correlated pairs, (2) validation, which requires a mapping from DCA pairs to a particular protein and the verification or analysis of the predicted contacts if a PDB structure is available, and (3) visualization of the DCA pairs as contact maps and the possibility to compare them against native contacts, as well as the depiction of predicted pairs with links in a 3D model

a given organism or to understand the influence of the number of sequences in the inference task. It is important to note that the alignments retrieved from Pfam are processed within our implementation `dca.m` to remove inserts (lower case letters) and the symbol "." that represents the spacing created by the inserts, since these inserts are not aligned by the profile HMM. Note also that the before mentioned reweighting takes care of almost repeated sequences, which carry little independent information. *See* **Note 1** in Subheading 4 for performance and input data requirements.

In addition to Fasta alignments from Pfam, the users can analyze their own datasets that might not be part of an HMM profile.

The only condition for applying `dca.m` is that data are aligned and provided in standard fasta format, with alignment gaps denoted by "-". In this chapter, we focus only on processing alignments from Pfam HMM.

3.1.2 Mean-Field DCA

Once the data is saved in a fasta-formatted file using a Pfam accession number or ID of interest, we will perform mean-field DCA on the data through the following steps:

1. Open Matlab and add the folder where the script `dca.m` is saved to the Matlab path. This can be done using the command `addpath('<folder path>')` or via the main Matlab window *File→ Set Path... → Add folder*
2. The script has a main function called:

```
dca(inputfile,outputfile)
```

The input parameters are as follows:

- `inputfile`—is a string listing the name of the file containing the MSA in fasta format (cf. below for the syntax).
- `Outputfile`—is a string listing the name of the output file. The output file is a text file with four columns and $L(L-1)/2$ rows, where L is the length of the Pfam domain (i.e., the number of MSA columns). Each row represents a residue pair and positions are listed in columns one and two; column four is the value of Direct Information, which is the main output of our mfDCA. Column three contains the value of Mutual Information, which is relevant for comparison only.
  Internal parameters:
- `theta`—is a parameter used to compute and assign a weight for sequences with certain degree of sequence identity. Sequences with an identity of (1 - `theta`) will be counted as redundant and have a smaller weight in the calculation. A default value of 0.2 was found to empirically provide good performance.
- `pseudocount_weight`—is a parameter optimized for a statistical correction done using pseudocounts. We have shown in ref. [7] that a default value around 0.5 is optimal for contact prediction.

A sample run of `dca.m` for the family of Uroporphyrinogen-III synthase HemD with Pfam Accession/ID: PF02602/HEM4, with $L=231$ and 3,110 sequences available, is shown below:

```
dca('HEM4.fasta', 'HEM4.DI')
```

The input file `HEM4.fasta`, obtained following the steps in the Data retrieval section, will produce the output file `HEM4.DI` that looks like this:

```
1   2   0.10758 0.131292
1   3   0.06264 0.007643
1   4   0.06653 0.033941
1   5   0.05926 0.006380
1   6   0.05649 0.006126
...
...
229 230 0.43428 0.298522
229 231 0.11028 0.035453
230 231 0.19088 0.160408
```

Columns one and two show the residue pairs, and the last column shows the computed DI value.

3. To obtain a list of domain pairs ranked by DI we just need to sort the output file (e.g., `HEM4.DI`) relative to the last column. This can be done easily using Unix bash commands and AWK. The following command will perform this task and create a new file with ranked domain contacts for residues with a sequence separation larger than four (less than four usually reflects trivial backbone couplings).

```
awk '$2-$1>4' HEM4.DI|sort -g -k 4 -r > HEM4_ranked.DI
```

The top ranked pairs, with a residue separation of at least four amino acids, look like this:

```
7   12  0.6539 0.2978
45  77  0.6810 0.2575
177 209 0.6444 0.2529
43  144 0.3808 0.2025
....
```

This shows that columns 7 and 12 are the highest directly coupled pair with a residue separation of at least four amino acids.

3.2 Validation and Analysis

Once we have a list of ranked pairs of directly coupled HMM columns we can use them for validation in a known protein crystal structure. This is done to verify that these are real contacts in a protein domain, or for analysis, to try to uncover biological roles or reasons of why these pairs are directly coupled. This section also describes how to predict specific protein contacts given a protein sequence that can be matched to the Pfam HMM of interest. To achieve this we perform the following protocol:

1. Download proteins of interest from the PDB. This can be done by visiting the following URL, http://www.pdb.org/, and by searching for the PDB ID of interest.
2. Given the knowledge that a specific protein in the PDB belongs to a Pfam family, we can map its sequence to the Pfam columns in the HMM. This mapping will allow us to uncover potential contacts in a specific PDB aligned to a Pfam family. A graphical view of this mapping can usually be found in the PDB site under Sequence tab→Annotations→Pfam. Using our previous example of the HEM4 family, we can obtain a map between the protein Uroporphyrinogen-III synthase (PDB ID: 1JR2) and the Pfam HMM profile of HEM4. To do this, we must have the protein sequence in fasta format and a copy of the sequence analysis tools developed by HMMER [23]. These tools can be downloaded at http://hmmer.janelia.org/software. For the case of HEM4, the following command using the `hmmscan` tool can be used:

```
hmmscan -o 1jr2_scan --notextw HEM4.hmm  1jr2.faa
```

`HEM4.hmm` is the HMM file that can be obtained from the Pfam Web site under "Curation & model", `1JR2.faa` is the protein sequence in fasta format and 1jr2_scan is the output file showing the alignment. To run `hmmscan` there is an intermediate step needed to compress and index the HMM model into a binary form to be read by `hmmscan`. This is done using this command:

```
hmmpress HEM4.hmm
```

This will generate files of the type `HEM4.hmm.h3*` which are accessed by `hmmscan` to perform the search. The file 1jr2_scan contains an alignment between the Pfam domain and the sequence in 1JR2.faa. For this particular example, we need to remove from 1JR2.faa, a leading sequence of amino acids (`'MGHHHHHHHHHHSSGHIEGRH'`) that is not numbered in the crystal structure. From this output we can extract an alignment similar to this:

```
family   family                                                    family
name     residue                                                   residue
  ↓        ↓                                                          ↓
HEM4     3 laaaleelGaeplelPlieieptedraeleaal [...] .kvdvvaeeptaeglv 230
1JR2    19 YIRELGLYGLEATLIPVLSFEF----LSLPSFS [...] aPVSCTAESPTPQALA 252
  ↑        ↑                        ↑              ↑                  ↑
protein  protein                   gaps          insert            protein
name     resid                                                     resid
```

This is telling us that residue 3 of the HMM is mapped to residue 19 on the protein sequence file. Gaps (-) are just sections of the HMM that have no corresponding mapping to the protein and the inserts are residues in the protein without a mapping to the HMM.

The information found in this alignment can be easily combined with the ranked domain residue pairs obtained before (e.g., `HEM4_ranked.DI`) to produce a list of the top N DI ranked protein residue pairs for a given protein. Such N pairs are the residue–residue couplets with the highest probability to be physical contacts in the protein. For more details about `hmmscan` or the use of hmm files please refer to the documentation in http://hmmer.janelia.org/software.

3. Once the list of DCA pairs has been produced for a given protein, it is possible to compute residue–residue distances to verify if such predicted pairs are physically close in the experimental structure. We can use these distances to calculate True Positive (TP) rates for the top N pairs to evaluate the quality of the predictions.

3.3 Visualization

We have developed tools to visualize the contacts predicted by DCA. One tool uses the list of amino acid pairs to create a two-dimensional contact map. These pairs come from the inference output of DCA (as shown at the end of the inference Subheading 3.1). The second tool creates a list of files and commands to be used by the Chimera molecular visualization system. These commands will render the contact predictions as links in a 3D representation of a protein when available. The following protocol describes the use of these two tools.

1. Use the Matlab script `plotDCAmap.m` to plot a predicted contact map. Refer to **Note 3** in Subheading 4 for details on the dependencies needed to run this script properly. The main function of this script is:

```
[Mat] = plotDCAmap(dca_pairs,native_pairs, pRange, ranking, mirror)
```

The input parameters are as follows:

- dca_pairs—a two columns vector of top DCA pairs. For comparison with a crystal structure the family residue pairs should be re-indexed to be consistent with the Pfam domain mapping in the PDB protein (*see* **Note 2**).
- native_pairs—an optional parameter for comparison with the native contact map. The format is a two columns vector with residue pairs of the "native" experimental structure. If there is no native information available then the input is just a pair of empty brackets "[ ]".
- pRange—the range of the protein in the format [initial_resID final_resID]. This is needed when the Pfam domain does not cover the complete protein.
- ranking—a 0/1 flag that instructs the function to color the residue pairs according to the DI rank. A color bar will appear to reflect the rank color mapping.
- mirror—a 0/1 flag to decide if we want to display the contact map also in the upper triangular part of the map.

One sample run for the protein BPTI would look like this:

```
[Mat] = plotDCAmap(BPTI_5pti_top80, CM_5pti_alpha, [1 58], 1, 0)
```

The first parameter is a Matlab variable (80 × 2 vector) created by importing a file containing the top 80 DCA pairs (one row per pair, one residue number per column) of the Kunitz_BPTI (Pfam: PF00014) family mapped to the bovine pancreatic trypsin inhibitor (58 amino acids, PDB ID: 5PTI). The second parameter has the C-alpha contact map of protein PDB: 5PTI. The third parameter is describing the protein range from residue 1–58, the next parameter is a flag telling the script to color the pairs based on their DI rank and the last parameter is not relevant for this example because we are comparing two maps, therefore we cannot generate a mirror image of the DCA map. The results are shown in Fig. 2. The native pairs are plotted in the upper triangular part of the map and the DI pairs in the lower triangular part. The color scheme is dependent on the ranking (this is optional) and the native contacts are always colored blue. The coloring options can be modified in the code to fit particular needs. A sample file with the Matlab variables can be downloaded from the DCA Web site along with the scripts described in this section.

2. To show predicted contacts on a 3D molecular model, we wrote a script that generates Chimera input files based on a list of residue pairs. The script is very simple and just reformats the input pairs to be read in Chimera. The bash script

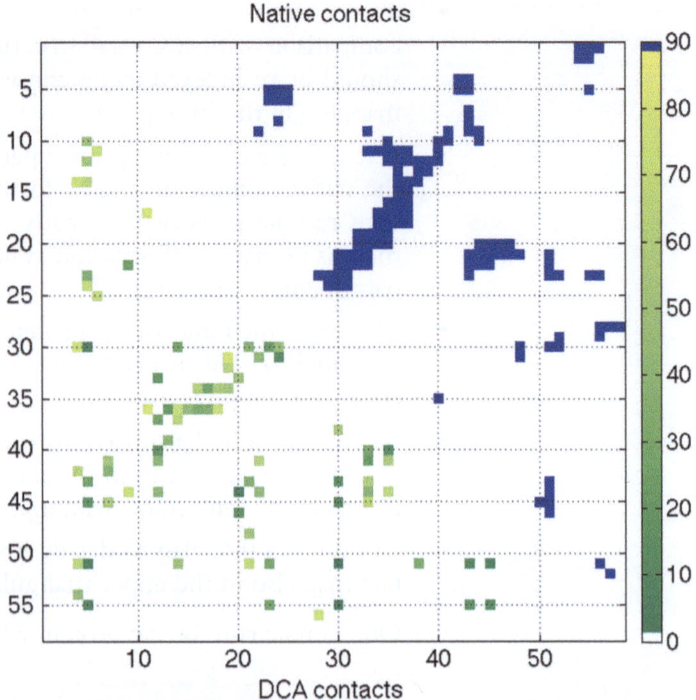

Fig. 2 An example of a contact map for the bovine pancreatic trypsin inhibitor (PDB ID: 5PTI) created with the top 80 DCA coupled pairs (*lower triangular region*) compared to the native C-alpha contacts (*upper triangular region*). The colorbar gradient corresponds to the DI rank with the extremes being the background (*white*) and the native contacts (*blue*). This figure was created with the Matlab script `plotDCAmap.m`

`GeneratePseudobonds.bash` has four input parameters: (1) An offset for the residue indexing in case the residue numbering does not coincide with the one in the 3D model, (2) the name of the text file with the contacts we want to display (two column format); in this file each pair uses one row and each residue number is separated by a space, (3) the filename prefix which is a user identifier for the output file, and (4) the chain ID to be used to display the bonds. This is especially important when we have models with more than one chain.

```
bash GeneratePseudobonds.bash offset pairs.map filename_prefix chain
```

The output of the script `GeneratePseudobonds.bash` is two files:

1. `pseudobond_<prefix>_dist<chain>_pairs.dat` which contains the residue pairs in a format that the pseudobond reader of Chimera will understand.

2. `pseudobond_<prefix>_dist<chain>_pairs.cmd` which instructs chimera to draw the pairs as links in a 3D model.

   Continuing the example for the trypsin inhibitor, we run the script:

```
bash GeneratePseudobonds.bash 0 5pti_dcaTop30.map 5pti A
```

In this example, there is no need for an offset hence the first parameter is 0. The second parameter is the file `5pti_dcaTop30.map` which has the top 30 DI ranked pairs for PDB 5PTI, the third parameter is just a prefix to be used to call the output files and the last parameter just refers to chain A of PDB 5PTI. Once the script has been run, there are two output files: `pseudobond_5pti_distA_pairs.dat` and `pseudobond_5pti_distA_pairs.cmd`.

Once the output files have been generated, we can open Chimera and load a 3D model, for example PDB ID: 5PTI. Then we can load the pseudobonds. To do that, we go to *Tools→ Depiction→ Pseudobond reader→ select file* `pseudobond_5pti_distA_pairs.dat`. The final step is to run the command to show the contacts, *File→ Open→ select file* `pseudobond_5pti_distA_pairs.cmd`. This will show the contacts as green links or bars in the 3D model. We can modify properties of the links, e.g., color or thickness of the link in *Tools→ General Control→ Pseudobond panel→ Attributes.*

Figure 3 shows the top 30 DCA contacts drawn as links in the 3D model of trypsin inhibitor (PDB ID: 5PTI). The inferred pairs are shown in green as links represented with sticks. These sample files can be downloaded with the scripts package in the DCA Web site.

3.4 Case Studies

Residue contacts estimated with DCA have been used in several studies to identify biologically relevant features of protein structure. In addition to making accurate predictions for intra-domain contacts, Morcos et al. showed that many of the directly coupled pairs that presented long intra-protein residue–residue distances were in fact oligomerization contacts [7]. Hence, long intra-domain distances became close contacts in homo-oligomeric systems as in the case of protein NtrC1 (PDB ID: 1NY6) of *Aquifex aeolicus*, which forms a ring like structure. Some of the highly correlated pairs were contacts that coevolved to keep the ring structure together.

Also in ref. [7], it is shown that some of the highly coupled pairs are a feature of distinct conformations of proteins, like in the case of the DNA binding domain of NarL (PDB ID 1JE8), which has active and inactive forms. This provides evidence that DCA uncovers structural features that are related to dynamic properties

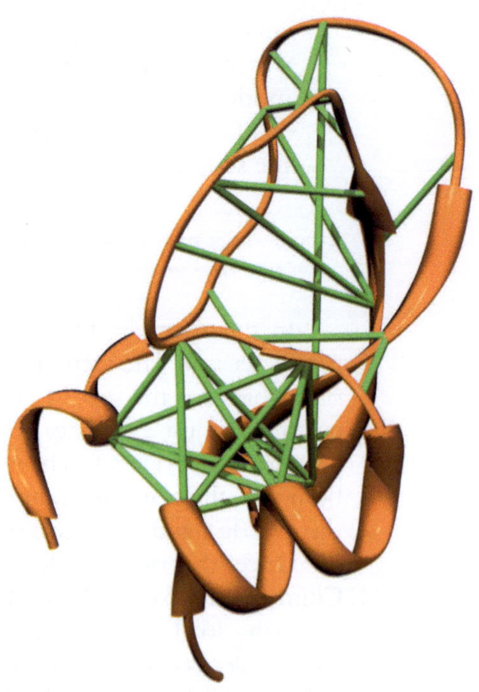

Fig. 3 The top 30 DCA contacts depicted as green sticks in the trypsin inhibitor (PDB ID: 5PTI). This figure was generated using Chimera with input files generated by the bash script `GeneratePseudobonds.bash`

of proteins. A similar example is also mentioned in ref. [19] for the case of the GlpT protein in *E. coli*.

Contact estimates based on DCA and DI ranking also provide the core for recent developments in protein structure prediction. The methodology called DCAfold [17] uses DCA contacts as parameters in a structure-based model (SBM) [25] in combination with accurate estimates of local information of a given protein. DCAfold generates structure predictions with average RMSD of 3.1 Å for proteins between 50 and 180 amino acids. The relevance of structure prediction using DCA contacts is also supported by more studies on structure prediction by Marks et al. [18] and a study on membrane protein structure prediction by Hopf et al. [19]. The use of methods to identify coevolving residues to attack the problem of protein structure estimation, including DCA, has been reviewed in ref. [26] and its importance for protein folding highlighted in ref. [27]. Finally, contact estimation with DCA has also been used in the study of protein interactions, for which coevolving residues provide a hint on binding interactions. One example of this is the case of the phospho-transfer systems observed in two component systems [14–16]. Coevolution of histidine kinases and response regulators left a signature that DCA was able to uncover.

4 Notes

1. *Inference.* The mfDCA script `dca.m` has an acceptable performance for families with up to 20,000 sequences. Processing larger families could affect performance. The script `dca.m` is also limited by the computing resources available. For an MSA with more than 20,000 sequences, the user could truncate the MSA to 20,000 or set a negative value in the internal parameter `theta`. This will turn off the reweighting procedure and reduce complexity at the expense of some predictive performance introduced by sequence bias. The protein length (L) is also constrained to a maximum range of 500–1,000 depending on the computer's memory. Longer proteins require the elimination of family columns to reduce computational complexity.
2. *Validation.* A correct mapping between the HMM to a protein of interest is required for proper contact estimation. Indexing mistakes or shifts could significantly affect performance.
3. *Visualization.* The Matlab script for contact visualization has a dependency when the contacts are colored by rank. It uses a modification of the HeatMap function to create customizable heatmaps. This version of the `heatmap` function can be found at : http://www.mathworks.com/matlabcentral/fileexchange/24253-customizable-heat-maps

Acknowledgments

This work was supported by the Center for Theoretical Biological Physics sponsored by the NSF (Grant PHY-0822283) and by NSF-MCB-1214457. JNO is a CPRIT Scholar in Cancer Research sponsored by the Cancer Prevention and Research Institute of Texas.

References

1. Göbel U, Sander C, Schneider R, Valencia A (1994) Correlated mutations and residue contacts in proteins. Proteins Struct Funct Genet 18:309–317
2. Lockless SW, Ranganathan R (1999) Evolutionarily conserved pathways of energetic connectivity in protein families. Science 286:295–299
3. Fariselli P, Casadio R (1999) A neural network based predictor of residue contacts in proteins. Protein Eng 12(1):15–21
4. Fariselli P, Olmea O, Valencia A, Casadio R (2001) Prediction of contact maps with neural networks and correlated mutations. Protein Eng 14(11):835–843
5. Pollastri G, Baldi P (2002) Prediction of contact maps by GIOHMMs and recurrent neural networks using lateral propagation from all four cardinal corners. Bioinformatics 18 Suppl 1:S62–S70
6. Hamilton N, Burrage K, Ragan MA, Huber T (2004) Protein contact prediction using

patterns of correlation. Proteins Struct Funct Bioinformatics 56(4):679–684
7. Morcos F et al (2011) Direct-coupling analysis of residue coevolution captures native contacts across many protein families. Proc Natl Acad Sci USA 108(49):E1293–E1301
8. Lunt B et al (2010) Inference of direct residue contacts in two-component signaling. Methods Enzymol 471:17–41
9. Weigt M, White RA, Szurmant H, Hoch JA, Hwa T (2009) Identification of direct residue contacts in protein-protein interaction by message passing. Proc Natl Acad Sci USA 106:67–72
10. Burger L, van Nimwegen E (2010) Disentangling direct from indirect co-evolution of residues in protein alignments. PLoS Comput Biol 6:e1000633
11. Taylor WR, Sadowski MI (2011) Structural constraints on the covariance matrix derived from multiple aligned protein sequences. PLoS One 6(12):e28265
12. Balakrishnan S, Kamisetty H, Carbonell JG, Lee SI, Langmead CJ (2011) Learning generative models for protein fold families. Proteins 79(4):1061–1078
13. Jones DT, Buchan DW, Cozzetto D, Pontil M (2012) PSICOV: precise structural contact prediction using sparse inverse covariance estimation on large multiple sequence alignments. Bioinformatics 28(2):184–190
14. Dago AE et al (2012) Structural basis of histidine kinase autophosphorylation deduced by integrating genomics, molecular dynamics, and mutagenesis. Proc Natl Acad Sci USA 109(26):E1733–1742
15. Schug A, Weigt M, Onuchic JN, Hwa T, Szurmant H (2009) High-resolution protein complexes from integrating genomic information with molecular simulation. Proc Natl Acad Sci USA 106:22124–22129
16. Schug A, Weigt M, Hoch J, Onuchic J (2010) Computational modeling of phosphotransfer complexes in two-component signaling. Methods Enzymol 471:43–58
17. Sulkowska JI, Morcos F, Weigt M, Hwa T, Onuchic JN (2012) Genomics-aided structure prediction. Proc Natl Acad Sci USA 109(26): 10340–10345
18. Marks DS et al (2011) Protein 3D structure computed from evolutionary sequence variation. PLoS One 6:e28766
19. Hopf TA et al (2012) Three-dimensional structures of membrane proteins from genomic sequencing. Cell 149(7):1607–1621
20. Nugent T, Jones DT (2012) Accurate de novo structure prediction of large transmembrane protein domains using fragment-assembly and correlated mutation analysis. Proc Natl Acad Sci USA 109(24):E1540–E1547
21. Finn RD et al (2010) The Pfam protein families database. Nucleic Acids Res 38:D211–D222
22. Berman HM et al (2000) The Protein Data Bank. Nucleic Acids Res 28:235–242
23. Eddy SR (2011) Accelerated profile HMM searches. PLoS Comput Biol 7(10):e1002195
24. Pettersen EF et al (2004) UCSF Chimera—a visualization system for exploratory research and analysis. J Comput Chem 25:1605–1612
25. Clementi C, Nymeyer H, Onuchic JN (2000) Topological and energetic factors: what determines the structural details of the transition state ensemble and "en-route" intermediates for protein folding? An investigation for small globular proteins. J Mol Biol 298:937–953
26. Marks DS, Hopf TA, Sander C (2012) Protein structure prediction from sequence variation. Nat Biotechnol 30(11):1072–1080
27. Dill KA, MacCallum JL (2012) The protein-folding problem, 50 years on. Science 338(6110):1042–1046

Chapter 6

ITScorePro: An Efficient Scoring Program for Evaluating the Energy Scores of Protein Structures for Structure Prediction

Sheng-You Huang and Xiaoqin Zou

Abstract

One important component in protein structure prediction is to evaluate the free energy of a given conformation. Given the enormous number of possible conformations for a sequence, it is extremely challenging to quickly and accurately score the energies of these conformations and predict a reasonable structure within a practical computational time. Here, we describe an efficient program for energy evaluation, referred to as ITScorePro (Copyright © 2012). The energy scoring function in the ITScorePro program is based on the distance-dependent, pairwise atomic potentials for protein structure prediction that we recently derived by using statistical mechanics principles (Huang and Zou, Proteins 79:2648–2661, 2011). ITScorePro is a stand-alone program and can also be easily implemented in other software suites for protein structure prediction.

Key words Protein structure prediction, Scoring function, Free energy, Statistical potentials, Knowledge-based

1 Introduction

One challenge in protein structure prediction is how to quickly and accurately assess the protein conformations generated by sampling algorithms [1–4]. No matter whether the modeling is template-based or ab initio, one common procedure is to sample a huge number of possible protein conformations based on a given sequence, followed by energy evaluation of each conformation [4]. Therefore, the availability of an efficient and reliable scoring function to evaluate the energies of an ensemble of structures and to rank the structures accordingly is valuable for improving the accuracy of a structure prediction program. Despite the achievements in the past three decades, scoring functions remain one of the bottlenecks in the protein structure prediction community [3, 4]. Approaches to scoring function development can be classified

Daisuke Kihara (ed.), *Protein Structure Prediction*, Methods in Molecular Biology, vol. 1137, DOI 10.1007/978-1-4939-0366-5_6, © Springer Science+Business Media New York 2014

into two broad categories: physics-based and knowledge-based. Physics-based approaches decompose the free energy into individual interaction energy components such as van der Waals interactions, electrostatic interactions, bond stretching, angle bending, and torsional forces, whose force field parameters are derived from quantum mechanics calculations [5–9]. Despite the clarity in the underlying physics and the success achieved, physics-based scoring functions are restricted by high computational cost when applied to the practice of protein structure prediction, particularly when many conformations are sampled. On the other hand, in knowledge-based approaches, empirical interaction potentials are extracted from the interatomic information embedded in known protein structures [10–12]. Despite the simple energy terms, knowledge-based scoring functions have captured some key features of the molecular interactions in each protein, and have been widely used for protein structure prediction [3].

Normally, there are two ways to derive the interaction potential parameters from known protein structures using knowledge-based methods. The first method uses advanced numerical optimization techniques, in which the potential parameters are optimized such that the native or near-native structures have lower energies than the decoys [13–21]. Because of the computational difficulty of high-dimensional optimization, the potentials are reduced to coarse-grained format, either contact-based or using a reduced protein representation. The second method for the extraction of pairwise potentials from native structures uses an inverse Boltzmann relation [22–25]. This approach is referred to as the Potential of Mean Force (PMF) approach. Since the pioneering work by Tanaka and Scheraga [10], the PMF approach has been widely used to develop distance-dependent or contact-based potentials at atomic or residue level for protein structure prediction.

A long-standing bottleneck in the PMF approach is the determination of the "reference state" in which there is no interaction between any two atoms/residues [22, 23, 26]. As described by Thomas and Dill, the ideal reference state is inaccessible because of atom connectivity, excluded volume, and other effects in proteins [22]. An inaccurate reference state would result in uncertainties of the derived potentials. Currently, most PMF scoring functions are based on a crude approximation of the reference state by randomly mixing all of the atoms in the training set of protein structures. Methods have been developed to improve the approximation of the reference state [27, 28].

To circumvent the challenging reference state problem, we have developed a statistical mechanics-based iterative method to extract distance-dependent, all-atom potentials for protein structure ranking and selection [29]. Instead of using the inverse Boltzmann relation from which the reference state problem arises, we derived the pair potentials iteratively by comparing the predicted pair distribution functions with the native global, physical pair distribution functions. The resulting scoring function is provided by the program—ITScorePro (Copyright © 2012). ITScorePro is a stand-alone program for fast evaluation of the energies of given multiple protein conformations. The program can be easily ported to other protein structure prediction packages as a reliable ranking/filtering scoring function or as one of the consensus scoring functions. The source codes and binary files of ITScorePro are freely available to academic users. To obtain an academic user license, please visit our website at http://zoulab.dalton.missouri.edu/software.htm. The details of how to use ITScorePro are described below.

2 Materials

2.1 Input Data

The only required input for ITScorePro is the three-dimensional protein structures in Protein Data Bank (PDB) [30] file format. The structures can be either determined by experimental methods or constructed by computational sampling algorithms. The program may take either one or multiple pdb files at a time. Each PDB file may contain one structure such as:

```
......
ATOM      1  N   PRO A   1      28.395  39.460   5.572  1.00 28.59           N
ATOM      2  CA  PRO A   1      29.435  38.616   4.975  1.00 29.18           C
ATOM      3  C   PRO A   1      28.923  37.839   3.767  1.00 28.03           C
ATOM      4  O   PRO A   1      27.748  37.920   3.416  1.00 27.72           O
ATOM      5  CB  PRO A   1      29.885  37.635   6.079  1.00 28.54           C
ATOM      6  CG  PRO A   1      28.634  37.557   6.962  1.00 29.49           C
ATOM      7  CD  PRO A   1      28.123  39.025   6.955  1.00 30.11           C
......
END
```

or multiple models in the PDB format of NMR structures such as:

```
......
MODEL        1
ATOM      1  N   PRO A   1      -7.620  14.411   3.801  1.00  0.00           N
ATOM      2  CA  PRO A   1      -6.833  14.660   5.048  1.00  0.00           C
ATOM      3  C   PRO A   1      -5.336  14.614   4.764  1.00  0.00           C
ATOM      4  O   PRO A   1      -4.870  13.841   3.949  1.00  0.00           O
ATOM      5  CB  PRO A   1      -7.172  13.559   6.053  1.00  0.00           C
ATOM      6  CG  PRO A   1      -8.192  12.644   5.391  1.00  0.00           C
ATOM      7  CD  PRO A   1      -8.572  13.286   4.060  1.00  0.00           C
......
ENDMDL
MODEL        2
ATOM      1  N   PRO A   1      -8.023  14.967   4.022  1.00  0.00           N
ATOM      2  CA  PRO A   1      -7.478  14.360   5.276  1.00  0.00           C
ATOM      3  C   PRO A   1      -5.981  14.088   5.127  1.00  0.00           C
ATOM      4  O   PRO A   1      -5.558  13.344   4.264  1.00  0.00           O
ATOM      5  CB  PRO A   1      -8.208  13.042   5.529  1.00  0.00           C
ATOM      6  CG  PRO A   1      -9.218  12.863   4.404  1.00  0.00           C
ATOM      7  CD  PRO A   1      -9.147  14.113   3.528  1.00  0.00           C
......
ENDMDL
```

```
MODEL        3
ATOM      1  N   PRO A   1      -7.359  14.357   4.045  1.00  0.00           N
ATOM      2  CA  PRO A   1      -6.434  14.207   5.209  1.00  0.00           C
ATOM      3  C   PRO A   1      -4.979  14.274   4.753  1.00  0.00           C
ATOM      4  O   PRO A   1      -4.580  13.615   3.812  1.00  0.00           O
ATOM      5  CB  PRO A   1      -6.680  12.841   5.847  1.00  0.00           C
ATOM      6  CG  PRO A   1      -7.792  12.164   5.057  1.00  0.00           C
ATOM      7  CD  PRO A   1      -8.238  13.149   3.973  1.00  0.00           C
......
ENDMDL
......
```

Two example pdb files are provided in the package for demos: 1AJV.pdb contains a crystal structure and 1BVE.pdb contains a set of NMR conformations.

2.2 Programs Included in the Software Package

The ITScorePro package includes three main files: `ITScorePro.for`, `potentials.dat`, and `README`.

2.2.1 ITScorePro.for

The file “ITScorePro.for” is the main program written in Fortran 90. It is a stand-alone program and does not depend on any other library. The source code can be compiled by using any Fortran 90 or newer compiler versions on any platform. The program can be easily implemented in other protein structure prediction software as an integrated part.

2.2.2 potentials.dat

The file “potentials.dat” provides the atomic pairwise interaction potentials of our knowledge-based scoring function, which were derived from the statistical mechanics-based iterative method. The potentials are based on 20 atom types grouped from the 167 heavy atoms of the 20 standard amino acids. Hydrogen-involved interactions are implicitly considered in our scoring function (*see* **Note 1**). The definitions of the 20 atom types are described in our published research article [29]. Unlike the van der Waals interaction energy which approaches positive infinity as the separation distance decreases to zero, the interaction potentials of ITScorePro are given a maximum penalty of +10 for the interatomic pairs within a cutoff distance. The finite penalty allows for limited atomic clashes in the protein structures so as to keep the structures that contain a few local distortions but are still physically reasonable. Such error tolerance is useful for fast sampling algorithms for protein structure prediction. Two examples of the pair potentials in “potentials.dat” are shown as follows:

```
......
   vvr( 5, 6,1:) = (/
*  10.0000, 10.0000, 10.0000, 10.0000,
*  10.0000,  5.2117,  4.1384,  3.6789,
*   3.4222,  3.1903,  2.8300,  2.1713,
*   1.4022,  0.6702,  0.1340, -0.2002,
*  -0.4121, -0.5724, -0.6725, -0.7079,
*  -0.6960, -0.6497, -0.5831, -0.5239,
```

```
*  -0.4449,-0.3615,-0.2884,-0.2208,
*  -0.1554,-0.1078,-0.0652,-0.0322,
*  -0.0079, 0.0197, 0.0205, 0.0331,
*   0.0248, 0.0247, 0.0135,-0.0078,
*  -0.0335,-0.0606,-0.0758,-0.0864,
*  -0.0863,-0.0979,-0.1018,-0.1054,
*  -0.1049,-0.0958,-0.0873,-0.0772,
*  -0.0690,-0.0665,-0.0639,-0.0617,
*  -0.0589,-0.0527,-0.0480,-0.0392,
*   0.0000, 0.0000 /)
......
   vvr(10,20,1:) = (/
*  10.0000,10.0000,10.0000,10.0000,
*  10.0000, 4.9225, 2.9445, 1.4825,
*   0.6436,-0.3471,-1.3166,-2.2078,
*  -2.8422,-3.0079,-2.7832,-2.2979,
*  -2.0627,-1.7968,-1.5246,-1.1893,
*  -0.9314,-0.8987,-1.0084,-1.0837,
*  -1.0704,-0.8744,-0.7414,-0.5940,
*  -0.4828,-0.3277,-0.2989,-0.1499,
*  -0.2185,-0.1834,-0.2329,-0.2024,
*  -0.1602,-0.1406,-0.0852,-0.0536,
*  -0.0324, 0.0067,-0.0109,-0.0269,
*  -0.0683,-0.0077, 0.0511, 0.0993,
*   0.0885, 0.0403, 0.0863, 0.1059,
*   0.1128, 0.0698, 0.0204, 0.0090,
*   0.0423, 0.0522, 0.0607, 0.0337,
*   0.0000, 0.0000 /)
......
```

Here, the array "vvr" is the name of the pair potentials. The first index i and second index j in "vvr (i, j, k)" stand for the atom types. Each atom type is assigned a number. The complete mapping between the atom names and atom type numbers are provided in the main program "ITScorePro.for." For example, atom type number 5 stands for the "aromatic carbons" and 6 for the "aliphatic carbons bonded to carbons or hydrogens only"; "vvr(5,6,1:)" are the interaction potentials between these two atom types at different separation distances. The third index k refers to the distance in terms of bins, with a bin size of 0.2 Å. For example, vvr(10,20,k), which is the kth value on the right-hand side of the equation in the second example, stands for the interaction potential between atom types 10 and 20 at distances from $(k-1/2)\times 0.2$ Å to $(k+1/2)\times 0.2$ Å. Specifically, vvr(10,20,1) = 10.000 for any interatomic distance ranging from 0.1 Å to 0.3 Å, vvr(10,20,14) = −3.0079 for any distance ranging from 2.7 Å to 2.9 Å, and so on.

2.2.3 README

The file "README" contains instructions on how to compile the program from the source code, how to use the program to calculate the energy scores of protein structures/conformations, and how to run the demos. A ready-to-run executable that was compiled on a Linux platform is also included in the release.

2.3 Website

The academic license of the ITScorePro program can be downloaded from our website at http://zoulab.dalton.missouri.edu/software.htm. The program can be obtained via ftp.

3 Methods

3.1 How to Build the Program (Optional)

ITScorePro is a stand-alone program and can be installed on any platform. A Linux-based executable is included in the package so that Linux users can carry out energy calculations by running the executable directly. However, if one prefers to build his own executable or if a user works on a platform other than Linux, he may run the following compiling command by using a Fortran 90 compatible compiler:

```
f90 ITScorePro.for -o ITScorePro
```

Here, "f90" is the compiler command and can be replaced by the user's own Fortran compiler. "ITScorePro.for" is the only source code for building the program and the executable will be named as ITScorePro with the above command.

3.2 How to Run the Program

ITScorePro is a user-friendly program. To learn about its usage, simply type the program name at the command prompt:

```
./ITScorePro
```

The following usage information about the program will pop up on the computer screen:

```
This program calculates the ITScorePro scores of protein structures.

Reference: Huang, S.-Y., Zou, X. Proteins 79: 2648-2661, 2011.

USAGE: ITScorePro prot1.pdb [prot2.pdb [prot3.pdb [...]]]

The input pdb file(s) can include a single structure or multiple structures in NMR-style.

Examples:
    ITScorePro 1BVE.pdb > scores.dat
    ITScorePro 1AJV.pdb 1BVE.pdb > scores.dat
```

The program may take multiple pdb files that are separated by space(s) as the input. Each pdb file may contain one model or multiple models in the PDB format of NMR structures. In the NMR style, each structural model is recognized by the key words "MODEL" and "ENDMDL." An example to calculate the energy scores using ITScorePro is as follows:

```
ITScorePro 1AJV.pdb 1BVE.pdb
```

The two input pdb files (1AJV.pdb and 1BVE.pdb) from the Protein Data Bank [30] are provided in the program package. Users may use their own pdb file(s) as input files. The above command will calculate the energy scores of all the models in the two pdb files and will display the results on the computer screen as follows (*see* **Note 2**):

```
1AJV.pdb    MODEL     1    -8583.79
1BVE.pdb    MODEL     1    -7973.89
1BVE.pdb    MODEL     2    -7982.91
1BVE.pdb    MODEL     3    -8053.50
1BVE.pdb    MODEL     4    -8008.58
1BVE.pdb    MODEL     5    -8052.87
1BVE.pdb    MODEL     6    -8161.93
1BVE.pdb    MODEL     7    -7959.12
1BVE.pdb    MODEL     8    -8126.66
1BVE.pdb    MODEL     9    -8059.79
1BVE.pdb    MODEL    10    -8076.85
1BVE.pdb    MODEL    11    -7989.70
1BVE.pdb    MODEL    12    -7998.96
1BVE.pdb    MODEL    13    -8038.61
1BVE.pdb    MODEL    14    -8114.22
1BVE.pdb    MODEL    15    -8077.05
1BVE.pdb    MODEL    16    -7969.94
1BVE.pdb    MODEL    17    -7914.71
1BVE.pdb    MODEL    18    -8077.49
1BVE.pdb    MODEL    19    -8002.83
1BVE.pdb    MODEL    20    -8043.58
1BVE.pdb    MODEL    21    -8101.42
1BVE.pdb    MODEL    22    -8035.06
1BVE.pdb    MODEL    23    -7946.87
1BVE.pdb    MODEL    24    -7971.15
1BVE.pdb    MODEL    25    -8122.80
1BVE.pdb    MODEL    26    -7878.77
1BVE.pdb    MODEL    27    -8128.11
1BVE.pdb    MODEL    28    -8081.25
```

Here, the first column shows the name of the pdb file that contains the protein structure(s) to be evaluated. The second and third columns list the model number in the pdb file shown in the first column. The corresponding calculated energy score is given in the last column. In this example, 1AJV is a crystal structure which contains only one model, and 1BVE contains NMR structures of 28 models. Users may save the output in a file for further analysis by using the following command:

```
ITScorePro 1AJV.pdb 1BVE.pdb > scores.dat
```

4 Case Studies

4.1 Test Case 1

This test set is the high resolution (HR) decoy datasets prepared by Floudas and colleagues [16], which contains 148 non-homologous proteins with 500–1,600 high resolution (HR) decoys for each protein. The root-mean-square deviations (RMSDs) of most of the decoys are less than 6–7 Å. This set tests whether ITScorePro can distinguish between native/near-native structures and other conformations with low RMSDs, which is important for structural refinement. Figure 1 shows the success rate of ITScorePro on ranking the native structures as first among the decoys. The figure also shows the average Pearson correlation coefficients (CC) between the energy scores and the RMSDs of the decoys. Achieving high CC is desirable for a scoring function. For reference purposes, Figure 1 also shows the results of several other scoring functions that have been published, including ITScorePP [31], DFIRE [32], dDFIRE [33], DOPE [28], HRSC [16], PMF [29], HR [34], TE13 [14], HL [35], and LKF [36].

4.2 Test Case 2

This test set consists of the CASP8 server decoys that contain 123 proteins. The decoys were generated by the CASP8 servers and are available from the official site of CASP8 (http://predictioncenter.org/). Only the decoys with full-length prediction were considered. The residues that are found in the decoys but are absent in the native structures were deleted for comparability. Considering the fact that the experimental structures are not known during CASP competitions, the native structures were excluded in the test set. The set contains a total of 25,003 decoys, with an average of 203 decoys per protein. One performance metric for scoring functions is the score-RMSD correlation; the higher the correlation, the better the performance. Figure 2 shows the average score-RMSD correlation of the decoys for all the proteins [29]. The figure also shows the percentage of the cases with significant correlation (i.e., $CC > 0.8$). For reference purposes, Fig. 2 also shows the results from several other scoring functions including PMF [29], MODELLER/DOPE [28], DFIRE 2.0 [32], dDFIRE [33], and ITScorePP [31].

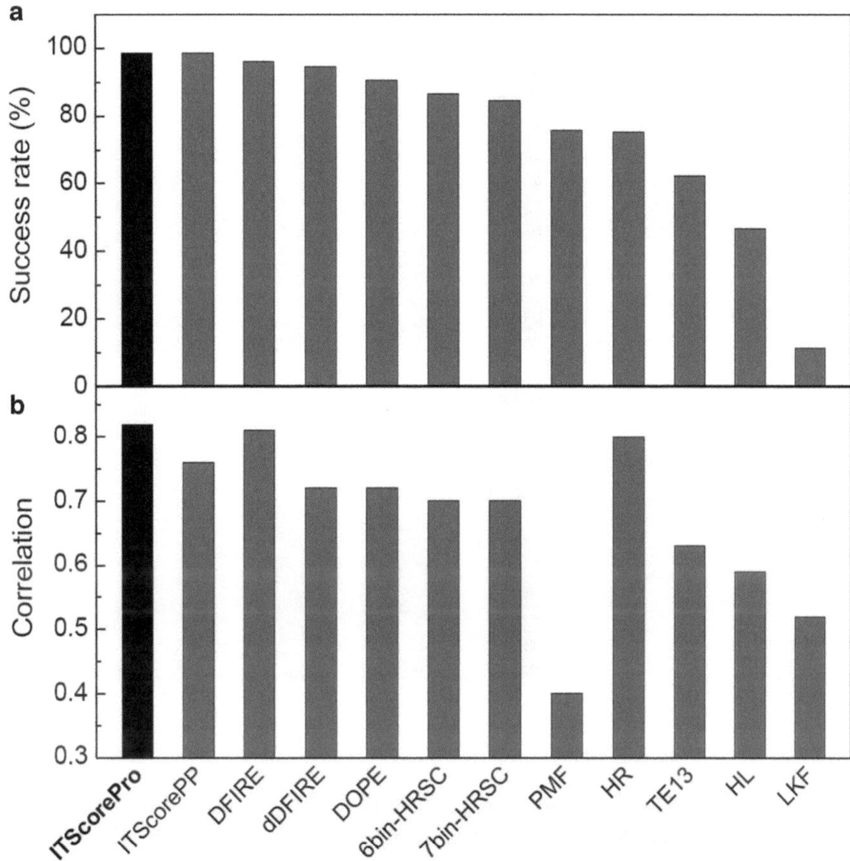

Fig. 1 (**a**) The success rate on recognizing the native structures and (**b**) average score-RMSD correlation coefficients for the high resolution (HR) decoy datasets of 148 proteins achieved by ITScorePro and 11 other scoring functions. PMF is a knowledge-based scoring function that we derived with an atom-randomized reference state for test purposes [29]

5 Notes

1. The scoring function of ITScorePro implicitly accounts for the effects of hydrogen atoms and ignores the hydrogen atoms during calculations. Therefore, it is unnecessary to add hydrogens before running ITScorePro and the presence of hydrogen atoms will not change the calculated energy score(s).
2. Because ITScorePro is based on the 20 standard amino acids, the scoring function does not recognize atom types from nonstandard amino acids and will ignore those heteroatomic records labeled by "HETATM ..." in the pdb file. This issue should be kept in mind when ITScorePro is used to calculate the energy scores of protein structures/conformations.

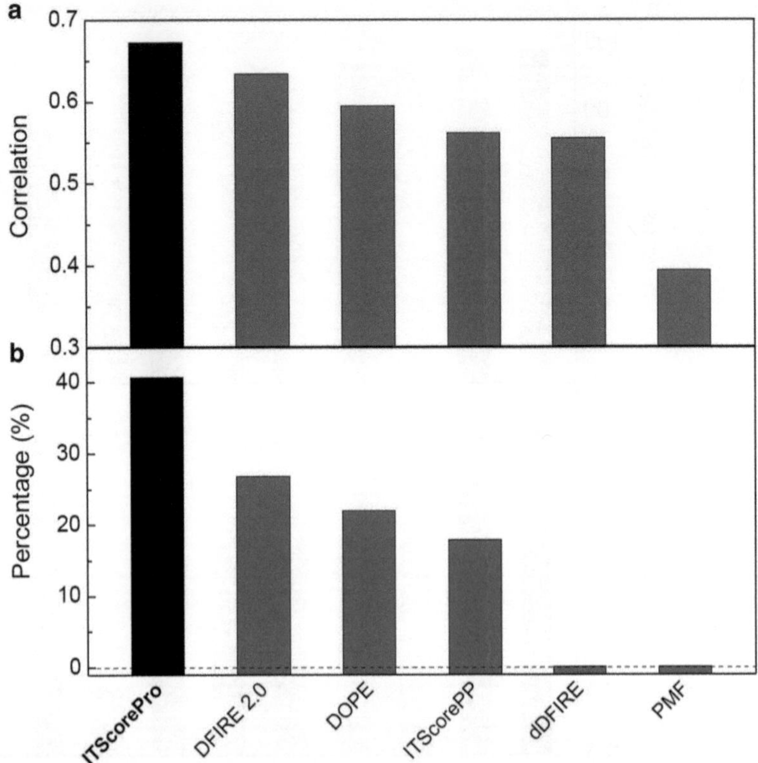

Fig. 2 (**a**) The average score-RMSD correlation and (**b**) the percentage of the cases with significant correlation (i.e., CC > 0.8) for the CASP8 decoy datasets of 123 proteins achieved by ITScorePro and several other scoring functions

Acknowledgments

X.Z. is supported by NIH grant R21GM088517, NSF CAREER Award DBI-0953839, the Research Board Award RB-07-32 and the Research Council Grant URC 09-004 of the University of Missouri. The computations were performed on the HPC resources at the University of Missouri Bioinformatics Consortium (UMBC).

References

1. Lazaridis T, Karplus M (2000) Effective energy functions for protein structure prediction. Curr Opin Struct Biol 10: 139–145
2. Buchete NV, Straub JE, Thirumalai D (2004) Development of novel statistical potentials for protein fold recognition. Curr Opin Struct Biol 14:225–232
3. Skolnick J (2006) In quest of an empirical potential for protein structure prediction. Curr Opin Struct Biol 16:166–171
4. Zhang Y (2008) Progress and challenges in protein structure prediction. Curr Opin Struct Biol 18:342–348
5. Brooks BR, Bruccoleri RE, Olafson BD, States DJ, Swaminathan S, Karplus M (1983) CHARMM: a program for macromolecular energy minimization and dynamic calculations. J Comput Chem 4:187–217
6. Wang J, Wolf RM, Caldwell JW, Kollman PA, Case DA (2004) Development and testing of a

general Amber force field. J Comput Chem 25:1157–1174
7. Case DA, Cheatham TE 3rd, Darden T, Gohlke H, Luo R, Merz KM Jr, Onufriev A, Simmerling C, Wang B, Woods RJ (2005) The Amber biomolecular simulation programs. J Comput Chem 26:1668–1688
8. Wroblewska L, Skolnick J (2007) Can a physics-based, all-atom potential find a protein's native structure among misfolded structures? I. Large scale AMBER benchmarking. J Comput Chem 28:2059–2066
9. Jagielska A, Wroblewska L, Skolnick J (2008) Protein model refinement using an optimized physics-based all-atom force field. Proc Natl Acad Sci USA 105:8268–8273
10. Tanaka S, Scheraga HA (1976) Medium-and long-range interaction parameters between amino acids for predicting three-dimensional structures of proteins. Macromolecules 9: 945–950
11. Miyazawa S, Jernigan RL (1985) Estimation of effective interresidue contact energies from protein crystal structures: quasi-chemical approximation. Macromolecules 18:534–552
12. Sippl MJ (1990) Calculation of conformational ensembles from potentials of mean force. An approach to the knowledge-based prediction of local structures in globular proteins. J Mol Biol 213:859–883
13. Maiorov VN, Crippen GM (1992) Contact potential that recognizes the correct folding of globular proteins. J Mol Biol 227:876–888
14. Tobi D, Elber R (2000) Distance-dependent, pair potential for protein folding: results from linear optimization. Proteins 41:40–46
15. Qiu J, Elber R (2005) Atomically detailed potentials to recognize native and approximate protein structures. Proteins 61:44–55
16. Rajgaria R, McAllister SR, Floudas CA (2008) Distance dependent centroid to centroid force. elds using high resolution decoys. Proteins 70: 950–970
17. Hao MH, Scheraga HA (1996) How optimization of potential function affects protein folding. Proc Natl Acad Sci USA 93:4984–4989
18. Bastolla U, Vendruscolo M, Knapp EW (2000) A statistical mechanical method to optimize energy functions for protein folding. Proc Natl Acad Sci USA 97:3977–3981
19. Mimy LA, Shakhnovich EI (1996) How to derive a protein folding potential? A new approach to an old problem. J Mol Biol 264:1164–1179
20. Huber T, Torda AE (1998) Protein fold recognition without Boltzmann statistics or explicit physical basis. Protein Sci 7:142–149
21. Koretke KK, Luthey-Schulten Z, Wolynes PG (1998) Self-consistently optimized energy functions for protein structure prediction by molecular dynamics. Proc Natl Acad Sci USA 95:2932–2937
22. Thomas PD, Dill KA (1996) Statistical potentials extracted from protein structures: how accurate are they? J Mol Biol 257:457–469
23. Koppensteiner WA, Sippl MJ (1998) Knowledge-based potentials—back to the roots. Biochemistry (Mosc) 63:247–252
24. Thomas PD, Dill KA (1996) An iterative method for extracting energy-like quantities from protein structures. Proc Natl Acad Sci USA 93:11628–11633
25. McQuarrie DA (2000) Statistical mechanics, University Science Books, Sausalito, California
26. Huang S-Y, Zou X (2010) Mean-force scoring functions for protein-ligand binding. Annu Rep Comput Chem 6:281–296
27. Zhou H, Zhou Y (2002) Distance-scaled, finite ideal-gas reference state improves structure-derived potentials of mean force for structure selection and stability prediction. Protein Sci 11:2714–2726
28. Shen M-Y, Sali A (2006) Statistical potential for assessment and prediction of protein structures. Protein Sci 15:2507–2524
29. Huang S-Y, Zou X (2011) Statistical mechanics-based method to extract atomic distance-dependent potentials from protein structures. Proteins 79:2648–2661
30. Berman HM, Westbrook J, Feng Z, Gilliland G, Bhat TN, Weissig H, Shindyalov IN, Bourne PE (2000) The Protein Data Bank. Nucleic Acids Res 28:235–242
31. Huang S-Y, Zou X (2008) An iterative knowledge-based scoring function for protein-protein recognition. Proteins 72: 557–579
32. Yang Y, Zhou Y (2008) Ab initio folding of terminal segments with secondary structures reveals the.ne difference between two closely related all-atom statistical energy functions. Protein Sci 17:1212–1219
33. Yang Y, Zhou Y (2008) Specific interactions for ab initio folding of protein terminal regions with secondary structures. Proteins 72: 793–803
34. Rajgaria R, McAllister SR, Floudas CA (2006) Development of a novel high resolution Ca-Ca distance dependent force field using a high quality decoy set. Proteins 65:726–741
35. Hinds DA, Levitt M (1994) Exploring conformational space with a simple lattice model for protein structure. J Mol Biol 243: 668–682
36. Loose C, Klepeis JL, Floudas CA (2004) A new pairwise folding potential based on improved decoy generation and side-chain packing. Proteins 54:303–314

Chapter 7

Assessing the Quality of Modelled 3D Protein Structures Using the ModFOLD Server

Daniel Barry Roche, Maria Teresa Buenavista, and Liam James McGuffin

Abstract

Model quality assessment programs (MQAPs) aim to assess the quality of modelled 3D protein structures. The provision of quality scores, describing both global and local (per-residue) accuracy are extremely important, as without quality scores we are unable to determine the usefulness of a 3D model for further computational and experimental wet lab studies.

Here, we briefly discuss protein tertiary structure prediction, along with the biennial Critical Assessment of Techniques for Protein Structure Prediction (CASP) competition and their key role in driving the field of protein model quality assessment methods (MQAPs). We also briefly discuss the top MQAPs from the previous CASP competitions. Additionally, we describe our downloadable and webserver-based model quality assessment methods: ModFOLD3, ModFOLDclust, ModFOLDclustQ, ModFOLDclust2, and IntFOLD-QA. We provide a practical step-by-step guide on using our downloadable and webserver-based tools and include examples of their application for improving tertiary structure prediction, ligand binding site residue prediction, and oligomer predictions.

Key words Model quality assessment, Protein tertiary structure prediction, Critical Assessment of Techniques for Protein Structure Prediction (CASP), Web servers, Single-model quality assessment methods, Consensus-based (clustering) model quality assessment methods, Per-residue error, Fold recognition, Ligand binding site residue prediction, Oligomer prediction

1 Introduction

Proteins are essential molecules in all living cells with numerous key functional and structural roles, both within and between cells [1, 2]. Since the advent of the CASP competition, a large number of template-based and template-free tertiary structure prediction methods have been developed, with the aim of producing 3D models of proteins from sequence. Routinely, these methods generate numerous 3D models with alternative conformations, but determining the most accurate conformation is challenging. Model quality assessment programs (MQAPs) are utilized for protein 3D

Daisuke Kihara (ed.), *Protein Structure Prediction*, Methods in Molecular Biology, vol. 1137, DOI 10.1007/978-1-4939-0366-5_7,

model quality prediction to help determine the most accurate 3D structural conformation. Hence, MQAPs have become a critical component in many of the leading protein tertiary structure prediction pipelines. Firstly, both the global and per-residue scores help in the estimation of how close a model might be to the native structure. Secondly, these scores provide details in regards to the potential errors within the model. Thirdly, the model quality scores give us a guide as to how useful the models will be in further computational and wet lab studies, such as improving tertiary structure prediction, ligand binding site residue prediction, oligomer predictions, molecular replacement, and mutagenesis studies [1–4].

In order to fully understand the necessity for and application of MQAPs, protein tertiary structure prediction methods are briefly discussed along with the CASP competition that has driven method development in this area. This is followed by a brief history of MQAPs, the various categories of methods including single-model and consensus-based, and a brief introduction to the practical use of our ModFOLD servers [5–7].

1.1 A Brief Introduction to Tertiary Structure Prediction

Protein tertiary structure prediction methods can be divided into two major subcategories, the purely template-based modelling (TBM) methods and those that are able to carry out template-free modelling (FM). Basically, if a structural template can be located in the PDB [8], then TBM methods such as homology modelling and fold recognition are utilized. However, if a structural template is unavailable then template-free modelling algorithms, which include physics-based methods and knowledge-based methods need to be utilized [2].

TBM is based on three key concepts: (1) Similar sequences fold into similar structures; (2) many unrelated sequences also fold into similar structures; and (3) there are only a relatively small number of unique folds when compared with the number of proteins found in nature; most of the fold space has been structurally annotated and few new folds are being solved [2, 9] (additionally *see* **Notes 1–4**).

Template-free modelling is also known as ab initio modelling, modelling from first principles, or de novo modelling. Template-free modelling is the prediction of protein tertiary structure from sequence, without utilizing a template protein structure. Template-free modelling involves the undertaking of conformational searches with the use of a designed energy function and results in the construction of several structural decoys based on potential conformations that will be utilized to select the final model. Template-free modelling energy functions are usually subcategorized into physics-based energy functions and knowledge-based energy functions, which are dependent on the utilization of statistics from experimentally solved protein structures [2, 10].

1.2 A Brief History of Model Quality Assessment

Since structural biologists first built theoretical protein models, algorithms have been developed to assess their quality. Early model quality methods were based on two broad concepts: assessing stereochemistry and predicting the free-energy of the model. Popular stereochemistry methods include PROCHECK [11], WHAT-CHECK [12], and a more recent method MolProbity [13]. These methods are mainly utilized to give a basic reality check of the constructed protein model. Nevertheless for multiple models determined as stereochemically correct, these methods are unable to judge among them. Additionally, stereochemical methods discard models with accurate backbone topology, which have stereochemically incorrect placement of side chains. Furthermore, stereochemical methods produce various alternative scores which cannot easily be summed up to produce a single score relating to overall model quality. Thus, these methods cannot truly be considered as MQAP methods in themselves; however some of the single-model MQAPs do contain several of these basic checks [3].

In addition, several physics-based methods have been developed for model quality assessment, which provide statistically determined energy functions, including ANOELA [14, 15] and DFIRE [16]. Alternatively, CHARMM [17] and AMBER [18] utilize numerous physics-based energy functions dependent on molecular force fields. Despite numerous attempts, the construction of a realistic energy function has remained a major challenge [2, 3].

1.3 Critical Assessment of Techniques for Protein Structure Prediction in Relation to Model Quality

The continuous development of more advanced protein structure prediction and model quality assessment tools is driven by the Critical Assessment of Techniques for Protein Structure Prediction (CASP) competition. CASP is a biennial competition whose main goal is the advancement of methods for the prediction of protein 3D structures from sequence. This is accomplished by providing objective testing of the methods via blind prediction. CASP is currently divided into numerous prediction categories, including: tertiary structure prediction—template-based and template-free modelling, disorder prediction, contact prediction, quality assessment, binding site prediction, and homo-oligomer prediction [2, 4, 19].

The quality assessment category (QA) was officially introduced in CASP7 (2006) with 28 methods participating [20]. The number of methods that competed in CASP9 (2010) had risen to 46 [21] and 37 server methods took part in the most recent CASP10 (2012) competition. The increasing number of independent quality assessment groups competing in CASP shows heightened interest for MQAP methods, which boosts competition and innovation in the field. Additionally, this demonstrates the critical role MQAPs now play in 3D modelling of proteins.

The CASP competition initially introduced two QA categories in CASP7, QMODE1 for global model quality prediction and

QMODE2 for per-residue quality prediction [20]. In CASP8 and CASP9 another assessment category emerged—QMODE3—where the per-residue errors from QMODE2 predictions are integrated into the B-factor column of the 3D models [4, 21]. One of the top QMODE3 prediction methods from the CASP9 competition was IntFOLD-QA/IntFOLD-TS [4, 22] (*see* **Note 3**).

1.4 Cutting-Edge Model Quality Assessment Methods

MQAPs are historically divided into two main categories: single-model-based methods that consider each model in isolation and clustering (or consensus)-based methods which compare multiple models for the same target. Single-model-based methods are comparable with consensus-based methods when a relatively small number of models are available. However, single-model methods currently lack accuracy when a wide range of models are available [5, 7], thus several groups have focused on their improvement [4, 23, 24]. In addition, there are methods and servers that blur the line between single-model methods and clustering approaches, which have recently been defined as the quasi-single-model methods [21]. Such quasi-single model approaches are able to provide accurate assessments of model quality given only a single model. They work by generating a number of alternative possible model conformations based on the target sequence, which are then compared with the target model using a clustering-based approach.

In contrast to single-model-based methods, consensus-based methods are often CPU intensive [7]. However, according to the previous two CASP experiments [21, 25], it has been found that clustering of numerous server models belonging to the same target results in the most accurate model quality predictions, both globally and on a per-residue basis [21, 25]. A list of the top MQAPs from CASP9 (2010) which have publicly available methods can be found in Table 1.

1.4.1 The ModFOLD 3.0 Server

The ModFOLD 3.0 server is a quasi-single-model-based server that implements IntFOLD-TS and ModFOLDclust2, therefore undertaking both single- and consensus-based model quality assessment. Our ModFOLDclust2 [7] (IntFOLD-QA [22]) consensus method (along with its predecessor ModFOLDclust) was amongst the top MQAPs that participated in the previous two CASP experiments (CASP8 and CASP9). The ModFOLDclust2 global quality score is composed of a simple linear combination of output scores from two methods, ModFOLDclust and ModFOLDclustQ. ModFOLDclust utilizes the TM-score structural-alignment scoring method [26] to compare a given model against several alternative models that have been constructed for a given protein target [7]. Whereas the ModFOLDclustQ method is a rapid, structural-alignment-free algorithm that utilizes an implementation of the Q-score [27] for multiple model comparison, rather than the time-consuming structural-alignment

Table 1
Publicly available model quality assessment methods that participated in CASP9

Method	Single/ consensus	Global/local scores	References	Web link
IntFOLD-QA	Single and consensus	Global and local scores	Roche et al. [22]	http://www.reading.ac.uk/bioinf/IntFOLD/
MetaMQAP	Single	Global and local scores	Pawloski et al. [48]	http://genesilico.pl/toolkit/mqap
ModFOLDclust2	Consensus	Global and local scores	McGuffin and Roche [7]	http://www.reading.ac.uk/bioinf/ModFOLD/
ModFOLDclustQ	Consensus	Global and local scores	McGuffin and Roche [7]	http://www.reading.ac.uk/bioinf/ModFOLD/
MUFOLD_WQA	Consensus	Global	Wang et al. [49]	–
MULTICOM	Consensus	Global and local scores	Cheng et al. [50]	http://sysbio.rnet.missouri.edu/multicom_toolbox/
ProQ	Single	Global and local scores	Larsson et al. [51]	http://www.sbc.su.se/~bjornw/ProQ/ProQ.cgi
QMEAN	Single	Global and local scores	Benkert et al. [52–54]	http://swissmodel.expasy.org/qmean/cgi/index.cgi
QMEANclust	Consensus	Global and local scores	Benkert et al. [52–54]	http://swissmodel.expasy.org/qmean/cgi/index.cgi

scores [7] (*see* Subheading 3). Furthermore, ModFOLDclust2, ModFOLDclust, and ModFOLDclustQ produce both global (QMODE1) and local/per-residue (QMODE2) quality scores. The per-residue errors produced by ModFOLDclust2 are amongst the most accurate and have subsequently been included in the B-factor column (QMODE3) of IntFOLD-TS 3D models [4] as part of the IntFOLD prediction pipeline [22]. (*See* Table 2 for a comparison of the methods and **Notes 1–6**).

The ModFOLD 3.0 server takes as input: an amino acid sequence, a set of 3D models (or a single model) for a given protein, a short name for the query sequence, and optionally an email address for the return of results. Figure 1 shows a screen capture of the ModFOLD 3.0 submission form. In Fig. 2 are the results of the ModFOLD 3.0 server using consensus mode (ModFOLDclust2) for an example CASP9 target (T0515). The machine-readable results can also be downloaded from the download link at the top of the main results page (Fig. 2). Additionally, the ModFOLD 3.0 server produces a model quality score (between 0 and 1—bad to good) and a *p*-value in relation to the confidence of the prediction, as can be seen in Fig. 2. The *p*-value confidence scores range from $P<0.001$ ("certain," colored blue) to $P<0.01$ ("high," colored

Table 2 Comparison of all of the ModFOLD server versions in terms of relative speed, upload options, output format, and method types

Method	Relative speed	Upload options	Output modes	Method type
ModFOLD v 1.1	Fast	Single/multiple models	QMODE1	Pure single-model
ModFOLD v 2.0	Slow	Single/multiple models	QMODE1, QMODE2, QMODE3	Quasi-single model
ModFOLD v 3.0 (Default mode)	Slow	Single/multiple models	QMODE1, QMODE2, QMODE3	Quasi-single model
ModFOLD v 3.0 (ModFOLDclustQ)	Fast	Multiple models only	QMODE1, QMODE2, QMODE3	Pure clustering
ModFOLD v 3.0 (ModFOLDclust2)	Slow	Multiple models only	QMODE1, QMODE2, QMODE3	Pure clustering
ModFOLD v 4.0 beta	Slow	Single/multiple models	QMODE1, QMODE2, QMODE3	Quasi-single model

green) to $P > 0.1$ ("poor" confidence, colored red) (*see* Fig. 2). The models are also colored by per-residue error using the same color scheme from blue to red (good to bad (Fig. 2)). Furthermore, the per-residue error plot (Fig. 3), highlights the residues of the predicted model with low confidence. This plot can additionally be downloaded as a PostScript file. Finally, clicking on the model in the main results page (Fig. 2) brings the user to a results page similar to Fig. 4 for the CASP9 target T0515. From the model results page (Fig. 4) the user can download the PDB file of the model with the per-residue errors in the B-factor column. Optionally, the Jmol Java applet may be deployed to display the model in 3D space. Version 4.0 of the ModFOLD server is also currently available for open beta testing.

If a user does not possess a set of models for analysis, then the IntFOLD server can be utilized (*see* **Notes 2–4**) which also integrates the ModFOLDclust2 method into a structure and function prediction pipeline [22]. The ModFOLD standalone software is available as downloadable Java executables (*see* Subheadings 2 and 3).

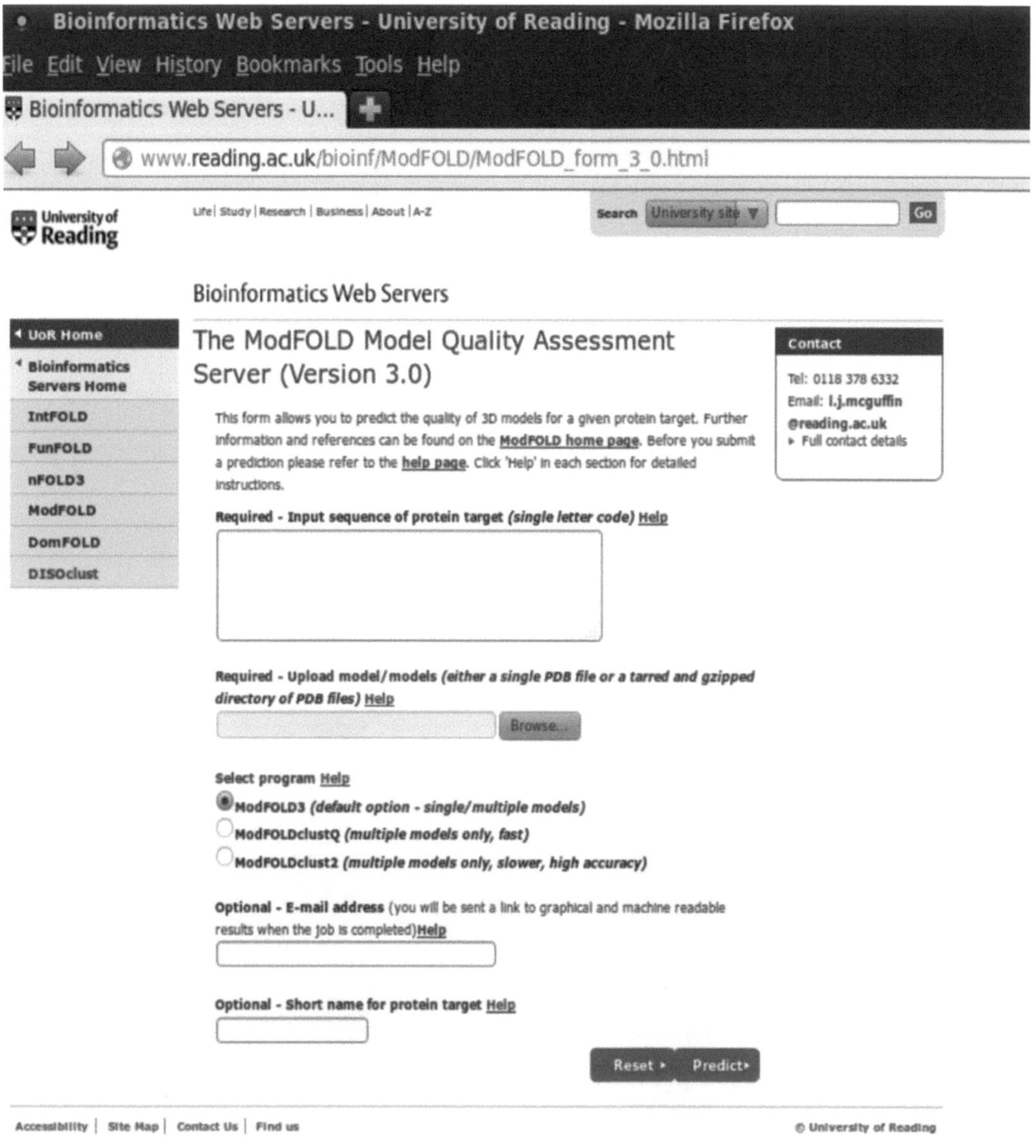

Fig. 1 Screenshot showing the ModFOLD 3.0 submission form. The web interface gives the user the opportunity to upload either single or multiple models for quality assessment

2 Materials and Systems Requirements

2.1 Web Server Requirements

For the model quality servers, such as ModFOLD 3.0, internet access, a web browser, a protein model (or a set of models), and the protein sequence are required. The servers are freely accessible at: http://www.reading.ac.uk/bioinf/ModFOLD/. Version 3.0 of the server additionally has the options of exclusively running

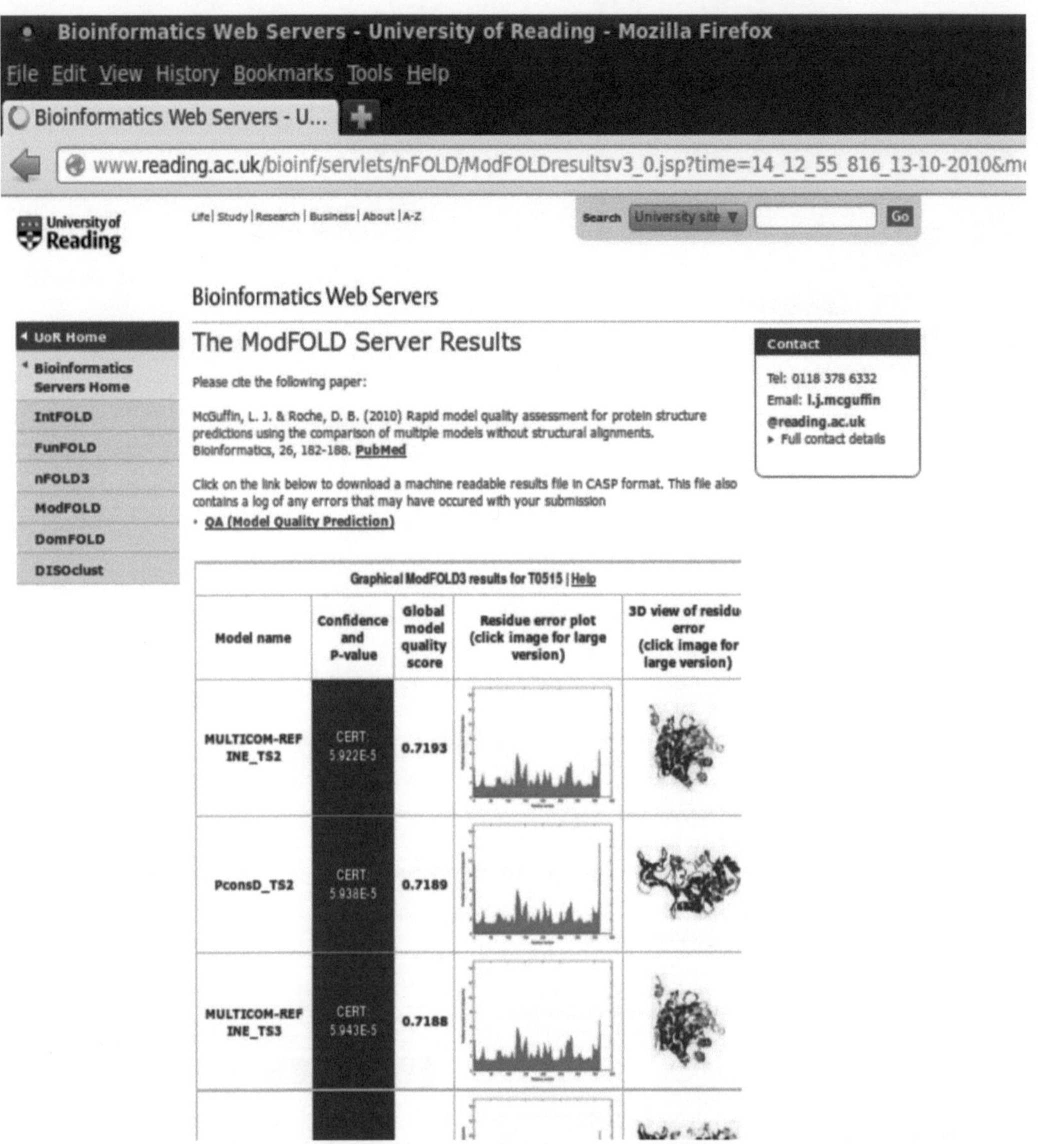

Fig. 2 Screenshot highlighting the ModFOLD 3.0 results page (in consensus mode—ModFOLDclust2) for the CASP9 target T0515. Machine-readable results can be downloaded from a link at the top of the page. Confidence *p*-values, model quality scores, and per-residue errors are provided for each model

either ModFOLDclustQ or ModFOLDclust2. *See* Table 2 for a list of available program versions and options and **Note 7** for common problems encountered.

2.2 Requirements for the Downloadable Executables

Downloadable executable versions of the ModFOLD component methods are available as executable JAR files which can be run locally. These executables have several dependencies and system requirements which are briefly described below for: ModFOLDclust,

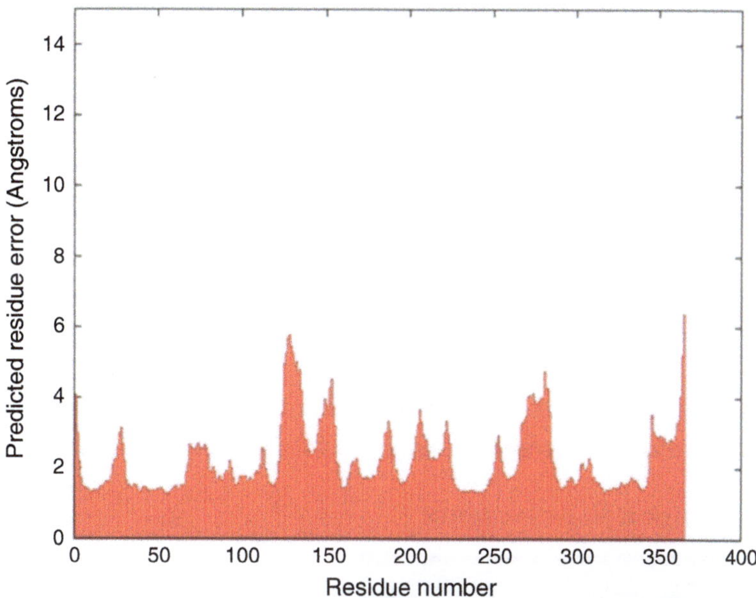

Fig. 3 Screenshot of the per-residue error plot results page for the top model from Fig. 2 (MULTICOM-REFINE) for the CASP9 target T0515. Additionally, the plot can be downloaded in PostScript format by clicking on the link at the bottom of the results page. This results page is accessed by clicking on the per-residue error plots in the main results page (Fig. 2)

ModFOLDclustQ, and ModFOLDclust2. The executables along with extensive README files and example input and output data can be downloaded from the following location: http://www.reading.ac.uk/bioinf/downloads/

2.2.1 ModFOLDclust

The system requirements are as follows:

1. A recent version of Java (java.com/getjava/).
2. Please ensure your system environment is set to English, as using other languages may cause problems with the ModFOLD clust calculations: export LC_ALL = en_US.utf-8.

2.2.2 ModFOLDclustQ

As ModFOLDclustQ is an alignment free score, the only system requirement is a recent version of Java (java.com/getjava/).

2.2.3 ModFOLDclust2

The system requirements are as follows:

1. A recent version of Java (java.com/getjava/).
2. Please ensure your system is running in English as using other languages may cause problems with the ModFOLDclust calculations: export LC_ALL = en_US.utf-8.

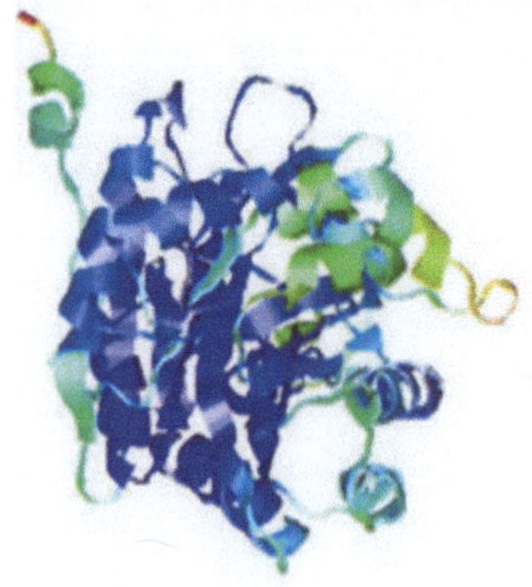

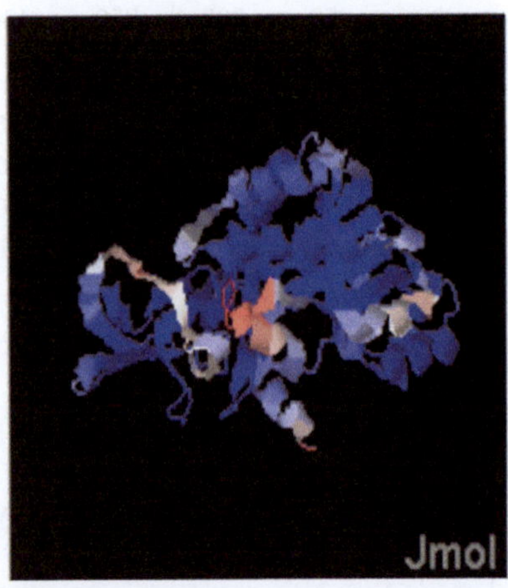

Fig. 4 Screenshot showing the results page for the top model from Fig. 2 (MULTICOM-REFINE) for the CASP9 target T0515. This page shows a large graphical representation of the model. A Jmol application allows users to examine the model in 3D space. Users can optionally download the PDB file of the model with the predicted per-residue errors in the B-factor column. Clicking on the models in the main results page (Fig. 2) brings users to this results page

3 Methods

The following is a step-by-step guide to generate model quality predictions using the latest web server implementations of ModFOLD.

3.1 Requisite Data for Servers

3.1.1 Sequence Data

Paste the full single-letter format amino acid sequence of the target protein into the appropriate text box (*see* Fig. 1). (Note, the sequence needs to be in FASTA format for use with the ModFOLDclust2 downloadable executables).

Sample sequence of CASP9 target T0515 (input in single-letter code):

```
MIETPYYLIDKAKLTRNMERIAHVREKSGAKALLALKC
FATWSVFDLMRDYMDGTTSSSLFEVRLGRERFGKETH
AYSVAYGDNEIDEVVSHADKIIFNSISQLERFADKA
AGIARGLRLNPQVSSSSFDLADPARPFSRLGEWDVPKVER
VMDRINGFMIHNNCENKDFGLFDRMLGEIEERFGALIARV
DWVSLGGGIHFTGDDYPVDAFSARLRAFSDRYGVQIYLE
PGEASITKSTTLEVTVLDTLYNGKNLAIVDSSIEAHMLD
LLIYRETAKVLPNEGSHSYMICGKSCLAGDVFGEFRFA
EELKVGDRISFQDAAGYTMVKKNWFNGVKMPAIAIRE
LDGSVRTVREFTYADYEQSLS
```

Ensure that the order of the residues in the query sequence corresponds to the sequence of residue coordinates in the model file. The server automatically renumbers the ATOM records in each model to match the residue position in the sequence. In cases where residues in the model file are not contained in the provided sequence, the quality prediction for the model will not be completed.

3.1.2 Model Data

Use the file selector to upload a PDB file of a model or multiple PDB files of models. Ensure that coordinates for each alternative model are contained within separate PDB files; single PDB files containing multiple alternative models will not be accepted. Multiple PDB files should be uploaded as a tarred and gzipped formatted archive file.

Steps to produce a tarball file for your own 3D models:

Linux/MacOS/Irix/Solaris/other Unix users

1. Tar up the directory containing your PDB files, e.g., type the following at the command line: tar cvf my_models.tar my_models/
2. Gzip the tar file, e.g., gzip my_models.tar
3. Upload the gzipped tar file (e.g., my_models.tar.gz) to the ModFOLD server.

Windows users

Use a free application such as 7-zip to tar and gzip the models.

1. Download, install, and run 7-zip.
2. Select the directory (folder) of model files to add to the .tar file and click "Add". Select the "tar" option as the "Archive format:" and save the file as something memorable, e.g., my_models.tar
3. Select the tar file and click "Add". Then select the "GZip" option as the "Archive format:"—the file should then be saved as my_models.tar.gz
4. Upload the gzipped tar file (e.g., my_models.tar.gz) to the ModFOLD server.

3.2 Choosing the Quality Assessment Program

Three program selectors are provided in ModFOLD version 3.0. In the ModFOLD 4.0 beta version only one default program option is currently provided (*see* **Note 8** on how to choose the best program for your requirements).

3.2.1 ModFOLD3

ModFOLD3 is the default method used for either single or multiple models. The method compares uploaded 3D models with those obtained from the IntFOLD-TS fold recognition method using the ModFOLDclust2 model quality assessment method (*see* Subheading 1.4.1).

3.2.2 ModFOLDclust2

ModFOLDclust2 is specific for multiple models. The slow but accurate clustering-based algorithm allows comparison of multiple models using structural alignments enabling the improved selection of target-template alignments. This method has been successfully used in multiple-template modelling as currently integrated into the IntFOLD2 server. This program option is best if there are multiple models for the target sequence, especially if the multiple models are built from alternative target-template alignments using several different methods.

The ModFOLDclust2 method can also work outside the web environment. It is provided in the form of an executable jar file (ModFOLDclust2.jar) and has been developed to run on Linux-based operating systems. This version of the program has been tested on recent versions of Ubuntu and CentOS, but it should work on most versions of Linux that have bash installed as long as the system requirements are met (*see* Subheading 2.2.3).

To run the program, edit the paths in the shell script (ModFOLDclust2.sh) and run. For example: **./ModFOLDclust2.sh T0515/home/liam/T0515.fasta /home/liam/T0515_example_models/**

Or follow the steps below.

1. Optionally, set the environment variable for Java, if Java has not been installed system-wide, e.g.

   ```
   export JAVA_HOME=/home/liam/jdk1.6.0/
   ```

2. Run ModFOLDclust2 with the target name, the sequence file (note that the sequence file should be in FASTA format, i.e., the header line should start with the > symbol with the single-letter amino acid sequence on the subsequent line(s)), and the models directory (note that multiple models of the target under analysis are required to produce model quality scores) included in the command. For example, if the target is "T0515", the sequence file is "/home/liam/T0515.fasta", and the models directory is "/home/liam/T0515_example_models/", then enter the following:

```
$JAVA_HOME/bin/java -jar ModFOLDclust2.jar
    T0515/home/liam/T0515.fasta/home/liam/
    T0515_example_models/
```

Otherwise, if you have java installed system-wide, enter:

```
java -jar ModFOLDclust2.jar T0515/home/liam/
    T0515.fasta/home/liam/T0515_example_models/
```

Ensure that the models are provided as separate files in PDB format and the sequence file in FASTA format. Note that FULL PATHS for your input file and models directory are required and that the models directory ends with a "/" (*see* **Note 7**).

A number of output files are produced in the models directory (e.g., "/home/liam/T0515_example_models/") and a log of the progress is printed to the screen as standard output. A description of the output files are as follows:

1. The QMODE2 output file—this file will consist of the target name plus "_ModFOLDclust2.out", e.g., "T0515_ModFOLDclust2.out". This file conforms to the CASP QA QMODE2 data format (http://predictioncenter.org/casp10/index.cgi?page=format#QA).
2. The sorted data file—this file will consist of the target name plus "_ModFOLDclust2.sort", e.g., "T0515_ModFOLDclust2.sort". This file contains the same data as the QMODE2 file but without the headers and in a more convenient machine-readable format.
3. B-factor files—these have the extension "*.bfact", e.g., "nFOLD3_TS1.bfact". These files contain your original model with the predicted per-residue error entered into the B-factor column. If you open these files using Pymol or Rasmol you can color your models according to the predicted errors with the b-factor/temperature coloring options.
4. Gnuplot files—these have the extension "*.gnuplot", e.g., "nFOLD3_TS1.gnuplot". These files contain per-residue error data for each model which can be plotted using gnuplot. The following is an example script:

```
set terminal postscript color
set output "nFOLD3_TS1.ps"
set boxwidth 1
set style fill solid 0.25 border
set ylabel "Predicted residue error (Angstroms)"
set xlabel "Residue number"
set yrange [0:15]
set yzeroaxis
unset key
```

```
set datafile missing "NaN"
plot "nFOLD3_TS1.gnuplot" using 1:2 with boxes,\
"nFOLD3_TS1.gnuplot" using 1:3 with points
quit
```

3.2.3 ModFOLDclustQ

ModFOLDclustQ is the quickest server option if there are several hundred models to compare. It uses a rapid clustering-based algorithm which does not require CPU intensive structural alignments. The program is also provided as an executable jar file and can be run in a similar way to the ModFOLDclust2 method described above.

3.3 Optional ModFOLD Server Inputs

Two optional text boxes can be filled in. One is for an email address to send links to the graphical or machine-readable results once the server has finished processing the data. The other is for assigning a preferred (memorable) short name for the prediction job that will enable the user to differentiate results returned by the server.

Acceptable job codes or names are restricted to the following set of characters: letters A-Z (either case), the numbers 0–9, and the following other characters: ., ~, _, - (excluding the commas). The job code or name specified is included in the subject line of the emailed link to results.

3.4 Server Fair Usage Policy

Standard web users are able to submit one job at any one time for each IP address. Once the first submitted job request is completed, notification of results is made via email, if email is provided. Alternatively, one can bookmark the job page where results can be found upon job completion. Once the job is completed, the IP address will be unlocked and the server is again ready for a new job request. The results of a completed job are retained on the server for 1 month.

3.5 Case Studies

The ModFOLD 3 server and the ModFOLDclust2 method have been used in several studies which have led to interesting biological findings. These studies are numerous with several examples centering on EFEO-cuperdoxins [42], *Schizosaccharomyces pombe* protein Translin [43], and Toll-like receptors [44, 45]. In addition, two recent studies have been undertaken in direct collaboration with the McGuffin group on areas of increased global research activity, namely cardiovascular disease [46] and food security [47].

The first case study focuses on food security in relation to *Blumeria graminis*, a plant pathogenetic fungi. This study combined proteogenomic and in silico structural and functional annotation to investigate the proteome of the pathogen [47]. Genome wide fold recognition was carried out using the IntFOLD server [22]. The quality of the models produced was then assessed using ModFOLD 3 [7], which resulted in several interesting conclusions in relation to the structural diversity of the genome. Firstly, the model quality assessment analysis showed that a large number of the models had low model quality, thus were probably novel folds or very distantly related to known structures. Secondly, six of the protein models had reasonably good model quality scores (greater than 0.4) and could be confidently assigned putative functions—glycosyl hydrolase activity—which has also been experimentally observed in previous wet lab studies. In conclusion, the model quality assessment helped to determine which protein models could confidently be assigned functions in this study and further highlighted the diversity of folds encoded by the *Blumeria graminis* genome [47].

The second case study is a more focused study on a specific protein kinase enzyme MST3, which has a putative role in cardiovascular development [46]. This study is experimentally based but used structure prediction (IntFOLD [22]) and model quality assessment (ModFOLDclust2 [7]) to help interpret the laboratory results. Basically, the laboratory results determined that the protein could not be entirely globular, but experimental results were unable to determine why. Modelling of the protein, followed by quality assessment and disordered prediction predicted that the enzyme has a large disordered domain, which was thought to be crucial to its function. The inclusion of the modelling and quality assessment in this study helped to explain the laboratory results and was crucial in proposing a new hypothesis of how this kinase-based pathway functions [46].

4 Notes

1. Model quality assessment methods such as ModFOLDclust2 [7] play an integral role in tertiary structure prediction and thus have been integrated into several prediction pipelines, including the IntFOLD server [22] for the prediction of protein structure, disorder, domain boundaries, and function form sequence. Furthermore, the integration of per-residue errors from ModFOLDclust2 [7] into the B-factor column of IntFOLD-TS [4] models presents the user with a guide to which parts of the model they can trust and which parts they cannot. Without such quality estimations it is difficult for a user to determine the usefulness of a generated 3D model [1, 4].

   Several of our recent tools for the prediction of structure and function from sequence have made extensive use of model quality prediction scores. Recently, the per-residue errors produced by ModFOLDclust2 [7] have been successfully utilized to guide multiple-template selection for improvement of our IntFOLD server models [34]. Furthermore, we recently developed FunFOLDQA [1], a novel quality assessment tool for protein–ligand binding site residue predictions. Finally, we are also testing a homo-multimeric (oligomer) prediction method, which makes use of ModFOLDclust2 scores in its prediction protocol. The above examples are not exhaustive, but highlight the integral and ubiquitous role model quality can play in structural bioinformatics. These example methods are briefly described below, followed by a discussion on common problems encountered in using MQAPs and reasons for choosing to use the web servers or the downloadable executables.

2. The IntFOLD server integrates numerous cutting-edge algorithms to predict protein structure and function from sequence. The IntFOLD server firstly utilizes numerous profile–profile alignment tools to produce 3D models for a target sequence. The ModFOLDclust2 quality assessment method is then used to rank the 3D models, producing both global and per-residue quality scores. The top five models in accordance with the global model quality scores become the output of IntFOLD-TS [4, 34]—the tertiary structure prediction component of the pipeline. Additionally, the per-residue errors from ModFOLDclust2 are added to the B-factor column of each model (*see* **Note 1** for details of the IntFOLD2-TS algorithm). The ModFOLDclust2 per-residue errors are also utilized by DISOclust 2.0 [35] to predict regions of disorder/high variability occurring in the protein. DISOclust 2.0 was one of the top disorder prediction methods in the CASP9 experiment [36]. In addition, the domain boundary prediction component of the IntFOLD pipeline, DomFOLD [22], utilizes the PDP method [37] to identify domain boundaries for the top IntFOLD-TS model ranked using ModFOLDclust2. Finally, FunFOLD [38], the ligand binding site residue prediction method, performs model-to-template superpositions of the top ranked 3D models (according to ModFOLDclust2) and related templates containing bound biologically relevant ligands, to identify potential binding site residues [38]. The FunFOLD method was one of the top 10 methods in the CASP9 FN prediction category [39].

3. The per-residue errors from ModFOLDclust2 [7] are included in the B-factor column of the IntFOLD-TS [4] 3D model files. The per-residue predictions are very useful for users to know which areas of the model they can trust and which areas of the model are less accurate. An absence of per-residue errors in generated models arguably makes them less useful for further study. The per-residue errors provided by ModFOLDclust2 and integrated into the IntFOLD-TS models were found to be amongst the most accurate by the CASP9 assessors of the QA (QMODE3) category [21].

4. Following on from our performance in the CASP9 QA per-residue error category [21] (*see* **Note 2**), our new TBM

method utilized the ModFOLDclust2 [7] per-residue errors predicted in single template models to construct improved models from multiple templates [34]. This method has been subsequently integrated into the IntFOLD annotation pipeline, now IntFOLD2, which is in its open beta testing phase. Additionally, the method was blind tested in the recent CASP10 prediction experiment.

5. The FunFOLDQA method [1] is a quality assessment tool for protein–ligand binding site residue prediction, which borrows many ideas from 3D model quality assessment methods. Currently, there is a lack of methods for assessing the quality of ligand binding site residue predictions prior to the availability of experimental data. Once experimental data is available, the predictions are usually assessed using both the MCC [40] and the BDT [41] scores as in the CASP9 official assessment [39]; however this requires experimentally solved 3D structures with bound ligands. Thus, FunFOLDQA was developed to assess ligand binding site quality, prior to the availability of experimental data. FunFOLDQA utilizes protein feature analysis in its assessment of quality, which includes structure-based feature scores and ligand-based feature scores. The FunFOLDQA algorithm was utilized to re-rank the FunFOLD predictions, which resulted in a statistically significant improvement [1].

6. A homo-multimeric manual prediction protocol was also tested in the recent CASP10. The multimeric prediction protocol made use of the ModFOLDclust2 [7] selected 3D server models, per-residue errors, and the lists of templates generated by our IntFOLD2 server [22, 34], with its multi-template [34] modelling protocol (*see* **Note 3**). The semiautomated protocol utilized in the CASP10 experiment will be subsequently automated and integrated into future versions of the IntFOLD server.

   Numerous examples of the roles that model quality prediction can play are outlined in **Notes 1–6**. Not only in 3D structure prediction per se but also in oligomer prediction, ligand binding site prediction, domain prediction, and disorder prediction. Model quality assessment now plays a large and critical role in the field of structural bioinformatics and thus needs to be considered an essential component of any 3D structure prediction algorithm and pipeline.

7. When using model quality assessment servers, several problems may be encountered. These mainly include, but are not limited to, the use of incorrect files. Each PDB file should include the coordinates for only one model; not a single PDB file containing the coordinates for multiple alternative models. For multiple models the coordinates should be uploaded as a tarred and gzipped directory of separate files.

The file format must be correct. All files should be uploaded as PDB files containing correctly formatted ATOM records. In addition, all PDB structures in a single submission should have been built for the same target, using the same target amino acid sequence.

For models uploaded to the ModFOLD 3 server in particular, the amino acid sequence submitted should correspond exactly to the amino acid sequence used to build the model(s)—not doing so is a common error. *See* Subheading 3 for more details on how to correctly use the servers and downloadable Java executables.

8. Table 2 (Subheading 1.4.1) shows that the ModFOLD server versions vary in relation to speed, input and output options, and sensitivity. As a general rule for quasi-single model and clustering methods, the more models (from alternative templates/alignments) you submit then the better the prediction results. A recommendation of 40 or more alternative models should achieve good cluster analysis. Alternatively, a sequence can be submitted to the IntFOLD2 server (*see* **Note 1**), which will build up to 90 alternative models, along with automatically predicting their global and per-residue model quality.

   When using the ModFOLD 3 server, the quality of the results you would like to obtain, the speed by which they are obtained, the output format of the results, and the required input need to be considered. For example, if you use the ModFOLDclustQ option, your results will be returned to you quickly (up to 150 times faster), but if you use the ModFOLDclust2 option, the response time is much slower but the results will be more sensitive [7]. Thus, the user needs to leverage response time for results with the quality of results obtained, when choosing which algorithm to utilize.

   Another consideration is the use of the web servers versus the downloadable Java applications. The ModFOLD web servers permit users to submit only one job at a time due to the server load balancing. If users would like to use the MQAPs frequently or for multiple models, for example analyzing many thousands of models, then we would recommend that users download and install the MQAPs for local execution. This gives the user freedom in the number of models which can be analyzed, provided they have adequate CPU capacity.

   For light (several predictions per week, less than or equal to 300 models per target) users of MQAPs, server submission is adequate; whereas for heavy users (20 or more predictions per week, greater than 300 models per target) the downloadable applications would be most useful. Extensive help pages are available for the ModFOLD 3 web server, and README files are available to help install and run the downloadable java applications.

Acknowledgments

University of Reading Faculty Studentship, MRC Harwell and the Diamond Light Source Ltd. (to. M.T.B.). This research leading to these results has received funding from the European Union Seventh Framework Programme (FP7/2007–2013) under grant agreement No. 246556 (to D.B.R.).

References

1. Roche DB, Buenavista MT, McGuffin LJ (2012) FunFOLDQA: a quality assessment tool for protein-ligand binding site residue predictions. PLoS One 7(5):e38219. doi:10.1371/journal.pone.0038219
2. Roche DB, Buenavista MT, McGuffin LJ (2012) Predicting protein structures and structural annotation of proteomes. In: Roberts GCK (ed) Encyclopedia of biophysics, vol 1. Springer, Berlin
3. McGuffin LJ (2010) Model quality prediction. In: Rangwala H, Karypis G (eds) Protein structure prediction: methods and algorithms. Wiley, New York, pp 323–342
4. McGuffin LJ, Roche DB (2011) Automated tertiary structure prediction with accurate local model quality assessment using the IntFOLD-TS method. Proteins 79 Suppl 10:137–146. doi:10.1002/prot.23120
5. McGuffin LJ (2007) Benchmarking consensus model quality assessment for protein fold recognition. BMC Bioinformatics 8:345. doi:10.1186/1471-2105-8-345
6. McGuffin LJ (2008) The ModFOLD server for the quality assessment of protein structural models. Bioinformatics 24(4):586–587. doi:10.1093/bioinformatics/btn014
7. McGuffin LJ, Roche DB (2010) Rapid model quality assessment for protein structure predictions using the comparison of multiple models without structural alignments. Bioinformatics 26(2):182–188. doi:10.1093/bioinformatics/btp629
8. Berman HM, Westbrook J, Feng Z, Gilliland G, Bhat TN, Weissig H, Shindyalov IN, Bourne PE (2000) The Protein Data Bank. Nucleic Acids Res 28(1):235–242
9. McGuffin LJ (2008) Protein fold recognition and threading. In: Schwede T, Peitsch MC (eds) Computational structural biology. World Scientific, London, pp 37–60
10. Lee J, Wu S, Zhang Y (2009) Ab initio protein structure prediction. In: Rigden DJ (ed) From protein structure to function with bioinformatics. Springer, London, pp 1–26
11. Laskowski RA, Moss DS, Thornton JM (1993) Main-chain bond lengths and bond angles in protein structures. J Mol Biol 231(4):1049–1067. doi:10.1006/jmbi.1993.1351
12. Hooft RW, Vriend G, Sander C, Abola EE (1996) Errors in protein structures. Nature 381(6580):272. doi:10.1038/381272a0
13. Davis IW, Murray LW, Richardson JS, Richardson DC (2004) MOLPROBITY: structure validation and all-atom contact analysis for nucleic acids and their complexes. Nucleic Acids Res 32(Web Server issue):W615–W619. doi:10.1093/nar/gkh398
14. Melo F, Devos D, Depiereux E, Feytmans E (1997) ANOLEA: a www server to assess protein structures. Proc Int Conf Intell Syst Mol Biol 5:187–190
15. Melo F, Feytmans E (1997) Novel knowledge-based mean force potential at atomic level. J Mol Biol 267(1):207–222. doi:10.1006/jmbi.1996.0868
16. Zhou H, Zhou Y (2002) Distance-scaled, finite ideal-gas reference state improves structure-derived potentials of mean force for structure selection and stability prediction. Protein Sci 11(11):2714–2726. doi:10.1110/ps.0217002
17. Brooks BR, Bruccoleri RE, Olafson BD, States DJ, Swaminathan S, Karplus M (1983) CHARMM: a program for macromolecular energy, minimization, and dynamics calculations. J Comput Chem 4(2):187–217. doi:10.1002/jcc.540040211
18. Weiner SJ, Kollman PA, Case DA, Singh UC, Ghio C, Alagona G, Profeta S, Weiner P (1984) A new force field for molecular mechanical simulation of nucleic acids and proteins. J Am Chem Soc 106(3):765–784. doi:10.1021/ja00315a051
19. Moult J, Fidelis K, Kryshtafovych A, Rost B, Tramontano A (2009) Critical assessment of

methods of protein structure prediction—round VIII. Proteins 77 Suppl 9:1–4. doi:10.1002/prot.22589

20. Cozzetto D, Kryshtafovych A, Ceriani M, Tramontano A (2007) Assessment of predictions in the model quality assessment category. Proteins 69 Suppl 8:175–183. doi:10.1002/prot.21669
21. Kryshtafovych A, Fidelis K, Tramontano A (2011) Evaluation of model quality predictions in CASP9. Proteins Struct Funct Bioinformatics (79 Suppl 10):96–106. doi:10.1002/prot.23180
22. Roche DB, Buenavista MT, Tetchner SJ, McGuffin LJ (2011) The IntFOLD server: an integrated web resource for protein fold recognition, 3D model quality assessment, intrinsic disorder prediction, domain prediction and ligand binding site prediction. Nucleic Acids Res 39(Web Server issue):W171–W176. doi:10.1093/nar/gkr184
23. Benkert P, Biasini M, Schwede T (2011) Toward the estimation of the absolute quality of individual protein structure models. Bioinformatics 27(3):343–350. doi:10.1093/bioinformatics/btq662
24. Kalman M, Ben-Tal N (2010) Quality assessment of protein model-structures using evolutionary conservation. Bioinformatics 26(10):1299–1307. doi:10.1093/bioinformatics/btq114
25. Cozzetto D, Kryshtafovych A, Tramontano A (2009) Evaluation of CASP8 model quality predictions. Proteins 77 Suppl 9:157–166. doi:10.1002/prot.22534
26. Zhang Y, Skolnick J (2004) Scoring function for automated assessment of protein structure template quality. Proteins 57(4):702–710. doi:10.1002/prot.20264
27. Ben-David M, Noivirt-Brik O, Paz A, Prilusky J, Sussman JL, Levy Y (2009) Assessment of CASP8 structure predictions for template free targets. Proteins 77 Suppl 9:50–65. doi:10.1002/prot.22591
28. Altschul SF, Madden TL, Schaffer AA, Zhang J, Zhang Z, Miller W, Lipman DJ (1997) Gapped BLAST and PSI-BLAST: a new generation of protein database search programs. Nucleic Acids Res 25(17):3389–3402
29. Jones DT, Swindells MB (2002) Getting the most from PSI-BLAST. Trends Biochem Sci 27(3):161–164
30. McGuffin LJ, Bryson K, Jones DT (2000) The PSIPRED protein structure prediction server. Bioinformatics 16(4):404–405
31. Soding J (2005) Protein homology detection by HMM-HMM comparison. Bioinformatics 21(7):951–960. doi:10.1093/bioinformatics/bti125
32. Remmert M, Biegert A, Hauser A, Soding J (2012) HHblits: lightning-fast iterative protein sequence searching by HMM-HMM alignment. Nat Methods 9(2):173–175. doi:10.1038/nmeth.1818
33. Eswar N, Webb B, Marti-Renom MA, Madhusudhan MS, Eramian D, Shen MY, Pieper U, Sali A (2006) Comparative protein structure modeling using Modeller. Curr Protoc Bioinformatics Chapter 5:Unit 5 6. doi:10.1002/0471250953.bi0506s15
34. Buenavista MT, Roche DB, McGuffin LJ (2012) Improvement of 3D protein models using multiple templates guided by single-template model quality assessment. Bioinformatics 28(14):1851–1857. doi:10.1093/bioinformatics/bts292
35. McGuffin LJ (2008) Intrinsic disorder prediction from the analysis of multiple protein fold recognition models. Bioinformatics 24(16):1798–1804. doi:10.1093/bioinformatics/btn326
36. Monastyrskyy B, Fidelis K, Moult J, Tramontano A, Kryshtafovych A (2011) Evaluation of disorder predictions in CASP9. Proteins 79 Suppl 10:107–118. doi:10.1002/prot.23161
37. Alexandrov N, Shindyalov I (2003) PDP: protein domain parser. Bioinformatics 19(3): 429–430
38. Roche DB, Tetchner SJ, McGuffin LJ (2011) FunFOLD: an improved automated method for the prediction of ligand binding residues using 3D models of proteins. BMC Bioinformatics 12:160. doi:10.1186/1471-2105-12-160
39. Schmidt T, Haas J, Gallo Cassarino T, Schwede T (2011) Assessment of ligand-binding residue predictions in CASP9. Proteins 79 Suppl 10:126–136. doi:10.1002/prot.23174
40. Matthews BW (1975) Comparison of the predicted and observed secondary structure of T4 phage lysozyme. Biochim Biophys Acta 405(2):442–451
41. Roche DB, Tetchner SJ, McGuffin LJ (2010) The binding site distance test score: a robust method for the assessment of predicted protein binding sites. Bioinformatics 26(22):2920–2921. doi:10.1093/bioinformatics/btq543
42. Rajasekaran MB, Nilapwar S, Andrews SC, Watson KA (2010) EfeO-cupredoxins: major new members of the cupredoxin superfamily with roles in bacterial iron transport. Biometals 23(1):1–17. doi:10.1007/s10534-009-9262-z

43. Eliahoo E, Ben Yosef R, Perez-Cano L, Fernandez-Recio J, Glaser F, Manor H (2010) Mapping of interaction sites of the Schizosaccharomyces pombe protein Translin with nucleic acids and proteins: a combined molecular genetics and bioinformatics study. Nucleic Acids Res 38(9):2975–2989. doi:10.1093/nar/gkp1230
44. Wei T, Gong J, Jamitzky F, Heckl WM, Stark RW, Rossle SC (2009) Homology modeling of human Toll-like receptors TLR7, 8, and 9 ligand-binding domains. Protein Sci 18(8):1684–1691. doi:10.1002/pro.186
45. Gong J, Wei T, Stark RW, Jamitzky F, Heckl WM, Anders HJ, Lech M, Rossle SC (2010) Inhibition of Toll-like receptors TLR4 and 7 signaling pathways by SIGIRR: a computational approach. J Struct Biol 169(3):323–330. doi:10.1016/j.jsb.2009.12.007
46. Fuller SJ, McGuffin LJ, Marshall AK, Giraldo A, Pikkarainen S, Clerk A, Sugden PH (2012) A novel non-canonical mechanism of regulation of MST3 (mammalian Sterile20-related kinase 3). Biochem J 442(3):595–610. doi:10.1042/BJ20112000
47. Bindschedler LV, McGuffin LJ, Burgis TA, Spanu PD, Cramer R (2011) Proteogenomics and in silico structural and functional annotation of the barley powdery mildew Blumeria graminis f. sp. hordei. Methods 54(4):432–441. doi:10.1016/j.ymeth.2011.03.006
48. Pawlowski M, Gajda MJ, Matlak R, Bujnicki JM (2008) MetaMQAP: a meta-server for the quality assessment of protein models. BMC Bioinformatics 9:403. doi:10.1186/1471-2105-9-403
49. Wang Q, Vantasin K, Xu D, Shang Y (2011) MUFOLD-WQA: a new selective consensus method for quality assessment in protein structure prediction. Proteins 79 Suppl 10:185–195. doi:10.1002/prot.23185
50. Cheng J, Li J, Wang Z, Eickholt J, Deng X (2012) The MULTICOM toolbox for protein structure prediction. BMC Bioinformatics 13:65. doi:10.1186/1471-2105-13-65
51. Larsson P, Skwark MJ, Wallner B, Elofsson A (2009) Assessment of global and local model quality in CASP8 using Pcons and ProQ. Proteins 77 Suppl 9:167–172. doi:10.1002/prot.22476
52. Benkert P, Kunzli M, Schwede T (2009) QMEAN server for protein model quality estimation. Nucleic Acids Res 37(Web Server issue):W510–W514. doi:10.1093/nar/gkp322
53. Benkert P, Schwede T, Tosatto SC (2009) QMEANclust: estimation of protein model quality by combining a composite scoring function with structural density information. BMC Struct Biol 9:35. doi:10.1186/1472-6807-9-35
54. Benkert P, Tosatto SC, Schomburg D (2008) QMEAN: a comprehensive scoring function for model quality assessment. Proteins 71(1):261–277. doi:10.1002/prot.21715

Chapter 8

3D-SURFER 2.0: Web Platform for Real-Time Search and Characterization of Protein Surfaces

Yi Xiong, Juan Esquivel-Rodriguez, Lee Sael, and Daisuke Kihara

Abstract

The increasing number of uncharacterized protein structures necessitates the development of computational approaches for function annotation using the protein tertiary structures. Protein structure database search is the basis of any structure-based functional elucidation of proteins. 3D-SURFER is a web platform for real-time protein surface comparison of a given protein structure against the entire PDB using 3D Zernike descriptors. It can smoothly navigate the protein structure space in real-time from one query structure to another. A major new feature of Release 2.0 is the ability to compare the protein surface of a single chain, a single domain, or a single complex against databases of protein chains, domains, complexes, or a combination of all three in the latest PDB. Additionally, two types of protein structures can now be compared: all-atom-surface and backbone-atom-surface. The server can also accept a batch job for a large number of database searches. Pockets in protein surfaces can be identified by VisGrid and LIGSITE^{csc}. The server is available at http://kiharalab.org/3d-surfer/.

Key words Structure-based function prediction, Protein surface, Surface shape comparison, Structure similarity, 3D Zernike descriptor, database search

1 Introduction

With the progress of structural genomics initiatives, an increasing proportion of solved protein structures are not functionally annotated. It remains a challenging task to understand the relationship between protein structure and function and to extrapolate a working mechanism of cellular machinery. To this end, a large number of computational approaches have been developed for function annotation based on protein sequences and their tertiary structures [1, 2]. Structure-based approaches are advantageous over sequence-based methods in the sense that distant relationships of proteins can be better identified by structure comparison, which can also often identify functional residues that are not localized in their amino acid sequences. Analogous to the important role of

Daisuke Kihara (ed.), *Protein Structure Prediction*, Methods in Molecular Biology, vol. 1137,
DOI 10.1007/978-1-4939-0366-5_8,

BLAST [3] in sequence-based function annotation methods, protein global structure searches against the Protein Data Bank (PDB) [4] are the basis of any structure-based functional elucidation of proteins.

An intuitive approach for comparing two protein structures is to align the atoms or residues of the proteins. However, the structural alignment procedures are time consuming, making it unfeasible for searching against the entire protein structure database in real-time. To speed up the database search, numerous alignment-free methods have been developed. For example, 3D-BLAST encodes the structure as a 1D sequence of alphabets [5]. The LightField Descriptor is constructed by 2D projections (a combination of 2D Zernike descriptors and Fourier coefficients) rendered from uniformly distributed points around a sphere that surrounds the protein surface [6]. Unlike the methods that are based on 1D or 2D representations, 3D moment-based shape representations truly capture the 3D geometrical shape of proteins. Among them, the 3D Zernike descriptor (3DZD) has been shown to be suitable for the efficient comparison of protein surfaces [7]. 3DZD is a series expansion of a 3D mathematical function and thus can represent a 3D object compactly as a vector of coefficients of the series function. Moreover, 3DZD is rotation invariant, which means that structure alignment is not needed prior to comparison. These two characteristics allow for fast, real-time searches of structure databases. 3DZD has been successfully applied for fast comparisons of various biological structure data [8] including ligand binding pockets [9, 10], low-resolution electron microscopy data [11], ligand molecules [12], and protein–protein docking [13, 14].

Here, we present 3D-SURFER 2.0, an upgraded web-based platform for high-throughput protein surface comparison, analysis, and visualization [15]. The server compares the protein surface of a single chain, a single domain, or a single complex against databases of protein chains, domains, complexes, or a combination of all three in the latest PDB. By using the 3DZD representation of the structures, the search process will be completed in a couple of seconds. A query structure can be selected by its PDB code or uploaded from a local computer of a user. In addition, local geometrical characteristics of a query protein such as pocket regions can be identified by VisGrid [16] or LIGSITE^{csc} [17]. Retrieved structures for a query are visualized with animations of rotating proteins. Clicking a protein icon in a search result will invoke another search from the clicked structure to surf into the protein structure space. Structures are associated with CATH codes [18] and conventional root mean square deviation (RMSD); main-chain structure alignment can be computed with the combinatorial extension (CE) algorithm [19].

2 Materials

2.1 Input Data

The input data of 3D-SURFER is a protein structure, which will be compared against a user-specified dataset of the entire PDB database. The input structure is provided by entering its identification (ID) code in the search window or by uploading a PDB format file to the server. The ID code of an input protein structure is named based on the PDB ID of the protein. If an entire structure (e.g., a protein complex) in a PDB file is chosen for input, the ID is the same as the PDB code (e.g., 7tim). A chain in a PDB file can be specified by adding a hyphen and the chain ID following the PDB code (e.g., 7tim-A). A domain of a chain can be specified by further adding a domain ID that is defined by CATH (e.g., 7tim-A-01). In each case, a search against the entire structure database is executed on-the-fly.

Two options are provided for protein surface representation: the surface is generated from all surface atoms or is constructed using only backbone atoms. This is because our previous work [11] showed that depending on the query structure, one of the surface representations will agree better in structure retrieval to the conventional fold classification (CE [19] and SCOP [20]). For a target database to be searched against, four types of datasets are prepared: protein single chains, domains, complexes, or a combination of the three. Additionally, the user can specify two types of filters: a CATH filter that avoids displaying multiple structures with the same CATH level and a length filter which retrieves only proteins whose lengths are similar to the query structure.

2.2 Programs Used in the Server

Individual programs of the server are seamlessly integrated behind the scenes and thus will not be visible to users of 3D-SURFER. Here, we briefly describe the programs and computational steps implemented within the server. The steps to calculate 3DZD are summarized as follows. First, the protein molecular surface is triangulated by MSROLL [21] (*see* **Note 1**). The extracted mesh is then discretized to generate a cubic grid (a binary voxelization, *see* **Note 2**). Finally, the 3DZD is computed [22] by taking the cubic grid as input (a vector of 121 invariants). An input protein structure represented as a vector can be compared with other structures by computing the Euclidean distance of the vectors (*see* **Note 3**). Moreover, the VisGrid and LIGSITE^{csc} programs are integrated for characterizing the geometry of local surface regions in a query structure. In addition, the conventional RMSD of Cα atoms can be computed between a query structure and retrieved structures specified by clicking the "RMSD" button. The CE program is invoked for the RMSD computation (*see* **Note 4**). The query structure is visualized with the Jmol applet (http://www.jmol.org/) (*see* **Note 5**).

2.3 Web Sites

3D-SURFER version 2.0 is available at http://kiharalab.org/3d-surfer/. The previous version is still accessible at http://kiharalab.org/3d-surfer/v1.0 or from a tab in the menu bar. The Web site has a navigation menu bar that includes tabs for a regular job submission, batch job submission, tutorial, reference, contact information, and 3D-SURFER version 1.0.

3 Methods

3.1 Submit a Query Entry

Submitting a query entry on the 3D-SURFER Web site consists of four steps. The details are explained below (Fig. 1).

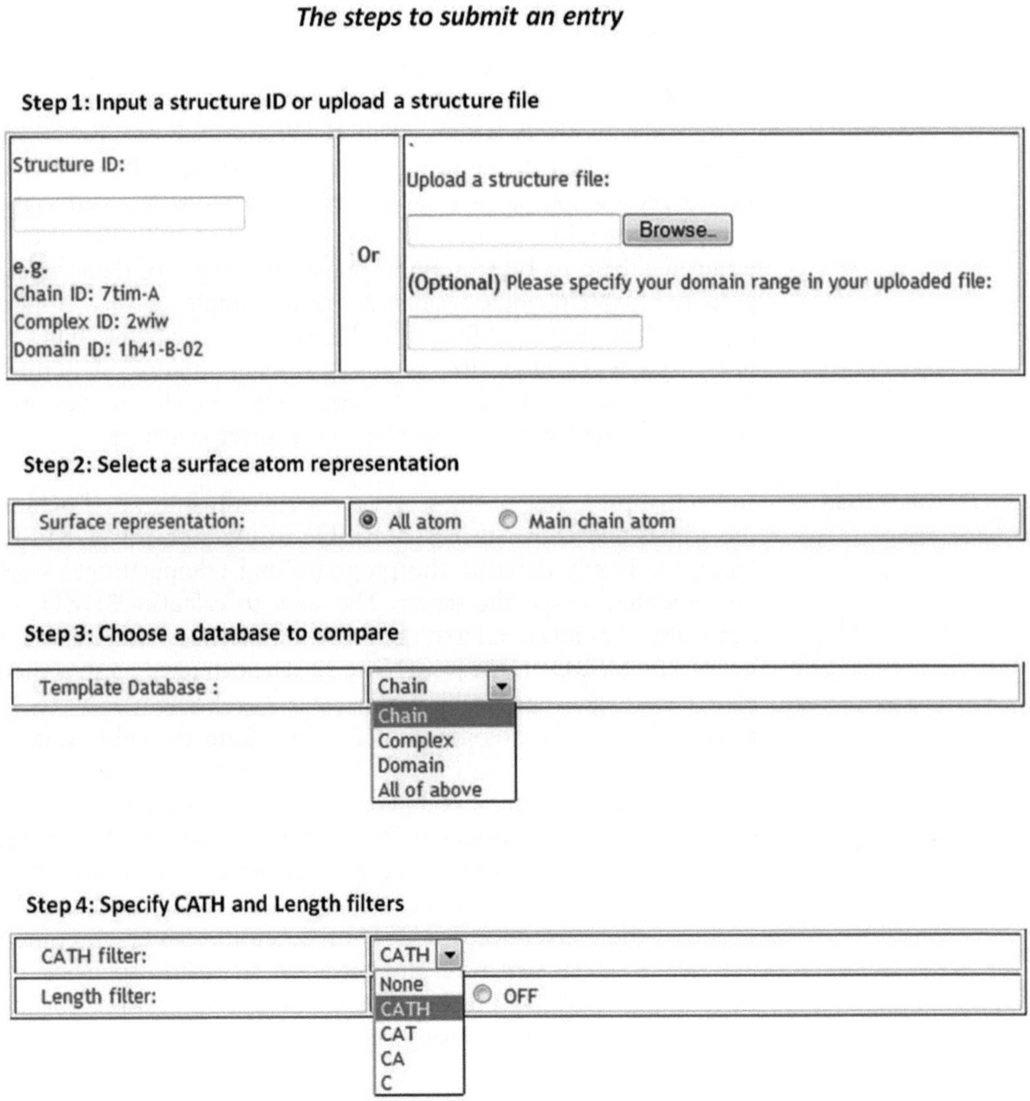

Fig. 1 The steps to submit a structure in 3D-SURFER

1. *Input a structure ID or upload a structure file*: In this step, users can either type a structure ID or upload a structure file. As shown in the example, the IDs should consist of four, six, or nine characters. When users type the first character, the text field will pop up a dropdown menu showing a list of IDs that start with the specified character. When the file upload option is used, the input should be consistent with the PDB format. At a minimum, the file should contain ATOM lines that contain atomic coordinates of proteins.
2. *Select a surface atom representation*: Two different surface representation options are provided: the all-atom-surface representation and the backbone-atom-surface representation, which includes Cα, C, and N atoms in the main chain. The choice should be made according to the purpose of the structure database search. If proteins with similar main-chain orientations are sought, the backbone-atom-surface representation works best in many cases (except for proteins with floppy tails and unpacked structures, *see* our previous work [11]).
3. *Choose a database to be searched*: Users can compare the surface of a single protein chain to a dataset of single chains, domains, complexes, or a combination of the three in the latest PDB.
4. *Specify CATH and length filters*: It is common that retrieved results contain many homologous proteins or essentially the same proteins in different conditions. Users can filter out such similar structures using the CATH filter. If users specify CATH filters ("CATH", "CAT", "CA", "C", or no filter) in the dropdown menu, the returned search results will contain only one structure from the same CATH classification of the specified level. For example, if the CATH filter level "CAT" is specified, no proteins in the retrieved list share the first three digits. When the length filter is turned on, the search result will only contain proteins with similar length to the query, i.e., those which have a length ratio between 0.57 and 1.75 to the query protein. Note that the size information of proteins is lost in 3DZD. However, our work [7] shows that it is uncommon for proteins with very different sizes to have the same surface shape and thus turning on the length filter may not drastically change the search results.

3.2 Result Page

In this section, we will show how the surface comparison results are presented. The similarity of two proteins are quantified by the Euclidean distance (the square root of the sum of the squares of the differences between corresponding values) between the 3DZD vectors (121 scalar values) of the proteins that represent protein surface shape. In the 3D-SURFER results panel, the Euclidean distance is shown at the label "EucD:". If two structures have an Euclidean distance less than a threshold of 10, they can be considered significantly similar.

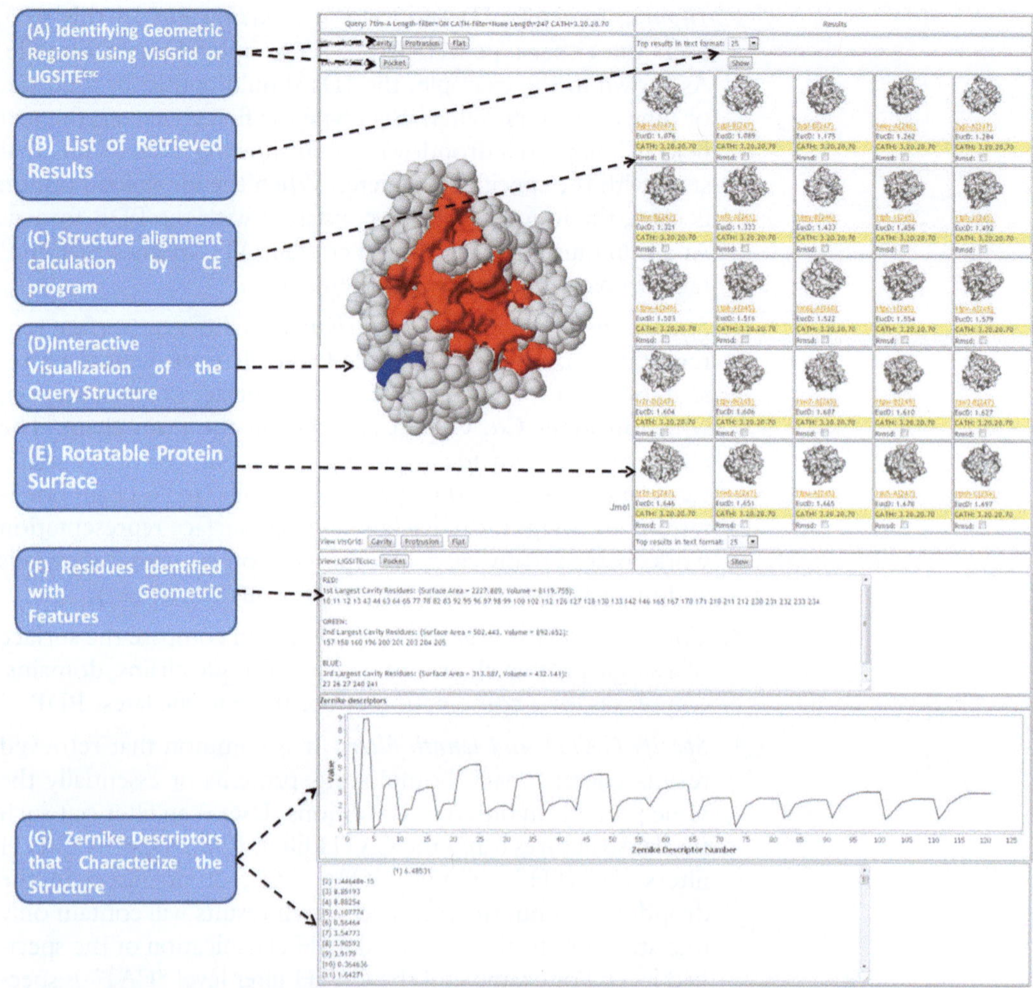

Fig. 2 The search result page of 3D-SURFER

1. *Viewing the query protein and its geometric features*: The query structure is visualized by the Jmol applet at the left side of the panel (Fig. 2d). By clicking the right button of the mouse, the representation of the protein can be changed using the functionality of Jmol. Geometrical features of the query structures, namely cavity, protrusion, and flat regions can be identified using VisGrid by clicking buttons located on the upper-left position of the panel (Fig. 2a). The "pocket" button in the third row invokes LIGSITE^{csc} to identify pockets on the protein surface. Identified pockets, protrusions, and flat regions are indicated with colors on the protein surface, where the rank is based on the size (*see* **Note 6**) (Fig. 2a, d, f). Users can

also look at the volumes and surface areas (*see* **Note 7**) of the convex hull (*see* **Note 8**) formed from the atom coordinates of the residues identified by VisGrid or LIGSITE^{csc}.

2. *Protein structure retrieval results and further navigation of the protein surface universe*: Retrieved structures from the database are shown on the right side of the panel. They are ranked by the Euclidean distance of the 3DZD. By default, the top 25 structures are displayed but the number of structures to show can be changed to the top 50, 100, 250, 500, or 1,000 (Fig. 2b). Protein surfaces can be rotated by moving the mouse over animation images of proteins. The images will spin 360° along both the *X* and *Y* axes to present a complete view of the protein surface (Fig. 2e). Each retrieved structure is associated with the PDB ID and the CATH code, if assigned (some PDB entries are not indexed in the CATH database). The PDB ID is linked to the entry in the PDB Web site.

   Clicking the image of a retrieved structure will invoke a new search from that structure against the database specified for the initial search. Welcome to the protein surface universe—users can enjoy surfing in the ocean of protein surfaces by taking advantage of real-time searches!

3. *Structure alignment calculations and visualization*: Conventional main-chain-based structure alignments can be performed using the CE program (*see* **Note 9**) to obtain RMSD values. When checking the "Rmsd:" box (Fig. 2c) of any retrieved similar protein, the CE program is invoked. Then the RMSD value is displayed and a new button will appear. By clicking this button, the structural alignment with the query structure is visualized using the Jmol applet on the left panel.

4. *3DZD invariants*: The 121 3DZD invariants of the query are shown in graphical form and in text (Fig. 2g).

3.3 Submit a Batch of Entries

When users want to benchmark 3D-SURFER by submitting many queries, they can use the batch mode page. When submitting a batch of query structures to 3D-SURFER, users can either type a custom list of structure IDs or upload a separate file with those IDs. Then, using the same steps for the submission of a single entry, users can go through the page to select a surface atom representation, a database to compare to, and specify the CATH and the Length filters. The extra step in this section is to select the number of entries to retrieve for each query. Taking 7tim-A as an example, Fig. 3 shows its retrieved results of the top 25 most similar structures.

```
*************************************************************************
Results for query: 7tim-A
*************************************************************************
RANK    Structure_ID     EUC_DIST        CATH            LENGTH
*************************************************************************
1       3ypi-A           1.076           3.20.20.70      247
2       1ypi-B           1.089           3.20.20.70      247
3       3ypi-B           1.175           3.20.20.70      247
4       1ney-A           1.262           3.20.20.70      246
5       2ypi-A           1.284           3.20.20.70      247
6       7tim-B           1.321           3.20.20.70      247
7       1nf0-A           1.333           3.20.20.70      261
8       1ney-B           1.433           3.20.20.70      246
9       1tph-1           1.456           3.20.20.70      245
10      1tph-2           1.492           3.20.20.70      245
11      1tpw-A           1.503           3.20.20.70      245
12      1tpb-1           1.516           3.20.20.70      245
13      1m6j-A           1.522           3.20.20.70      260
14      1tpc-1           1.554           3.20.20.70      245
15      1tpv-A           1.579           3.20.20.70      245
16      1r2r-D           1.604           3.20.20.70      247
17      1tpv-B           1.606           3.20.20.70      245
18      1sw7-A           1.607           3.20.20.70      245
19      1tpw-B           1.610           3.20.20.70      245
20      1sw3-B           1.627           3.20.20.70      247
21      1r2s-D           1.646           3.20.20.70      247
22      1sw0-A           1.651           3.20.20.70      247
23      1tpu-A           1.665           3.20.20.70      245
24      1su5-A           1.678           3.20.20.70      247
25      1tmh-C           1.697           3.20.20.70      254
*************************************************************************
```

Fig. 3 Retrieved 25 most similar structures for query 7tim-A. EUC_DIST is the Euclidean distance between the query 7tim-A and each retrieved structure. Length shows the number of amino acids residues in the proteins

4 Notes

1. *Surface generation*: The protein surface is constructed using the MSROLL program in Molecular Surface Package [23]. The surface consists of all the points of the van der Waals surface at which a spherical probe (i.e., the solvent radius of 1.4 Å) can touch it. The output of the program is in the form of a polyhedral surface triangulation. Surface computation with MSROLL is failed occasionally. In those cases, the MSMS program (http://mgltools.scripps.edu/) is invoked.
2. *Voxelization*: The constructed surface of a protein is mapped on a 3D grid, where voxels (cubic cells) that overlap with the

surface are marked with 1 and 0 otherwise (inside and outside of the protein). This discrete representation of the protein surface is used as an input for computing 3DZD.

3. *Euclidean distance*: The Euclidean distance d_E of two vectors X and Y is defined as:

$$d_E = \sqrt{\sum_{i=1}^{n} (X_i - Y_i)^2} \quad (1)$$

where n is the number of elements in the vectors. n is set to 121 in 3D-SURFER.

4. *Alignment by the CE program*: The CE program is used to align and compute the RMSD of two protein structures. The algorithm first identifies a set of structurally similar main-chain fragment pairs from the two protein structures and then assembles the fragment pairs to form a larger region of similar conformations. Thus, CE compares the main-chain conformations of two protein structures.

5. *Jmol applet*: Jmol is an open-source Java viewer for protein structures in 3D. Jmol can be integrated into web pages to display molecules in various representations. For example, molecules can be displayed as "ball and stick" models, "space filling" models, "ribbon" models, etc. To change the representation of a protein structure, right-click on the Jmol applet area, select the Surface menu, and then the sub-menu Molecular Surface option.

6. *Color of local regions identified by VisGrid or LIGSITE^{csc}*: The size of identified pockets and protrusions is ranked by colors. Red indicates the largest cavity/protrusion/pocket, green indicates the second largest, and blue indicates the third largest cavity/protrusion/pocket. Yellow indicates identified flat regions by VisGrid. The residues in the colored regions are provided in a window (Fig. 2f) on the results page.

7. *Volumes and surface area of a geometric region*: The Qhull program [24] is employed to calculate volumes and the surface area of the geometric regions identified. Qhull calculates volumes and surface areas by computing a convex hull of a specified local surface region (e.g., pocket).

8. *Convex hull*: The convex hull of a set Q of points is the smallest convex polygon P, for which each point in set Q is either on the boundary of P or in its interior. For example, the convex hull of 13 points, p_0 to p_{12}, in a two-dimensional space is given by $CH(Q) = \{p_0, p_1, p_3, p_{10}, p_{12}\}$ in Fig. 4.

9. *The CE alignment file corresponding to a RMSD calculation*: The RMSD between a retrieved structure and a query structure can be computed by checking the RMSD checkbox.

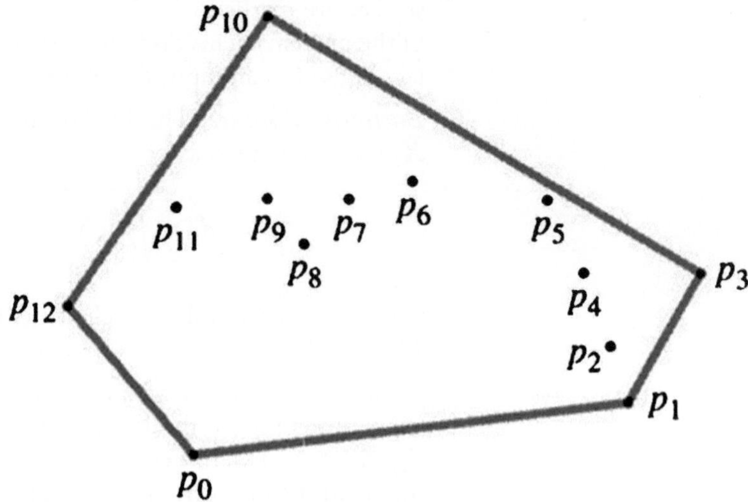

Fig. 4 An example of a convex hull for points in the 2D space

Then, a new button and the RMSD value will be shown. Clicking the RMSD button will invoke Jmol to visualize the structure superimposition between the retrieved structure and the query. The numerical value is a link to a text file that contains the CE alignment result file.

5 Case Studies

We show two examples of 3D-SURFER results. The first example demonstrates that 3DZD detects similar global surface shapes of proteins with a different overall fold and fold class (completely different secondary structure elements) (Fig. 5). In Fig. 5, the protein on the left is a β class protein (PDB code: 1 bar-A; fibroblast growth factor), while the one on the right is an α class chain (PDB code: 1rro-A; oncomodulin). The corresponding CATH IDs of the two chains are obviously different, 2.80.10.50 and 1.10.238.10, respectively. Despite the complete difference of the main-chain conformation of the proteins, their surface shapes exhibit surprising similarity with a 3DZD Euclidian distance of 12.66.

The new version of 3D-SURFER extended the coverage of structure databases to include protein domains and complexes, in addition to single chains. A search can be performed between different types of structure data, for example, a single protein chain can be compared against the shape of multi-chain protein complexes. Figure 6 shows a query single chain (PDB code: 2WHH-A) and a complex structure retrieved by a search, which is a homodimeric protein complex (PDB code: 2FDE). A close examination of

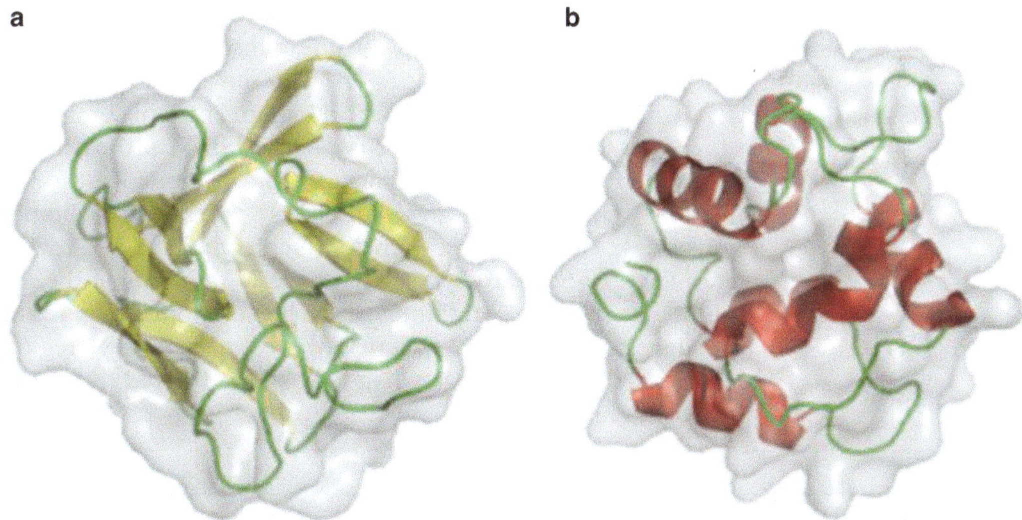

Fig. 5 An example of similar surface shapes of two proteins in different fold class. (**a**) 1 bar-A, a β-class protein; (**b**) 1rro-A, an α-class protein. (The figure is modified from Fig. 5 of ref. [7])

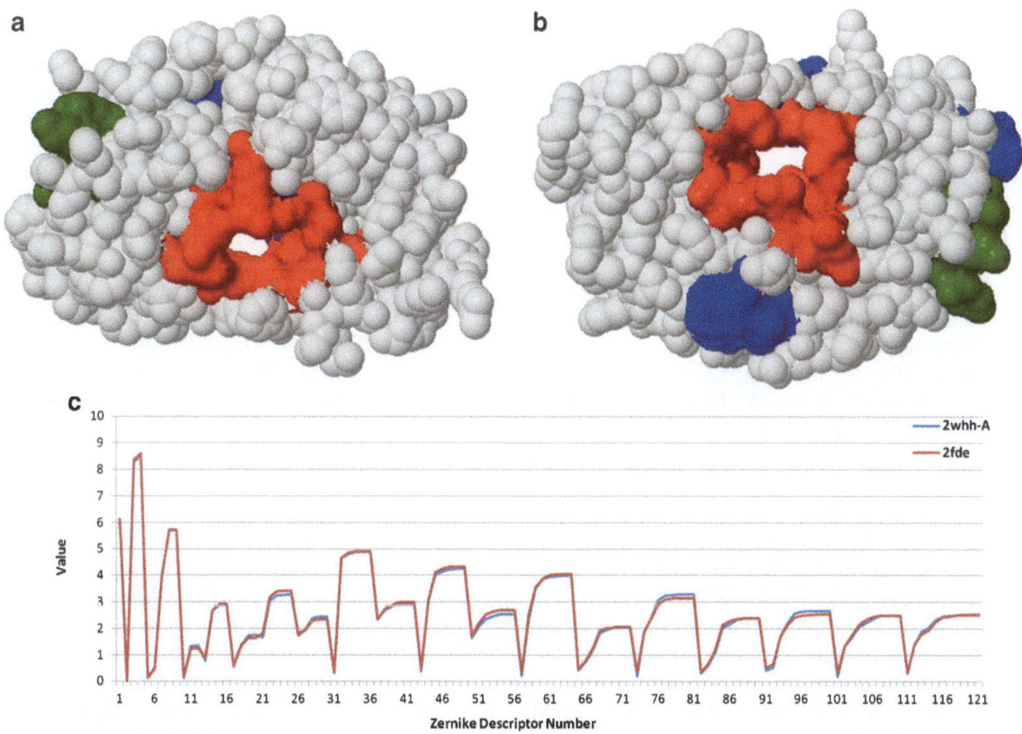

Fig. 6 An example of similar surface shapes of a single protein chain and a two-chain complex. (**a**) The protein chain 2WHH-A and its largest pocket (*in red color*) identified by LIGSITE^{csc}; (**b**) The complex 2FDE, comparable to 2WHH-A both in global shape and in pocket regions; (**c**) The 3DZD values of 2WHH-A and 2FDE

Fig. 6c shows that these two structures have nearly identical vectors of 3DZD invariants with a Euclidean distance of 0.982. Both of these are HIV-1 proteases [25, 26] with similar enzymatic properties, which are vital targets for the design of antiviral compounds in the treatment of AIDS. 2FDE is a native HIV-1 protease and a homodimeric enzyme in which the active site is located at the subunit interface. 2WHH-A is a mutant HIV-1 protease and a homologous single-chain tethered dimer, which contains a five residue linker, Gly-Gly-Ser-Ser-Gly, that links the N-terminus of the second monomer to the C-terminus of the first monomer. Our results also suggest that the two structures have similar pocket regions identified by LIGSITE^{csc} (shown in Fig. 6a, b). The red regions are pocket sites binding to inhibitors.

Acknowledgments

This work has been supported by grants from the National Institutes of Health (R01GM075004 and R01GM097528), National Science Foundation (EF0850009, DBI1262189, IOS1127027), and National Research Foundation of Korea Grant funded by the Korean Government (NRF-2011-220-C00004). The authors thank Kristen Johnson for proofreading the manuscript.

References

1. Sael L, Chitale M, Kihara D (2012) Structure- and sequence-based function prediction for non-homologous proteins. J Struct Funct Genomics 13(2):111–123
2. Hawkins T, Kihara D (2007) Function prediction of uncharacterized proteins. J Bioinform Comput Biol 5(1):1–30
3. Altschul SF, Madden TL, Schaffer AA, Zhang JH, Zhang Z, Miller W, Lipman DJ (1997) Gapped BLAST and PSI-BLAST: a new generation of protein database search programs. Nucleic Acids Res 25(17):3389–3402
4. Berman HM, Westbrook J, Feng Z, Gilliland G, Bhat TN, Weissig H, Shindyalov IN, Bourne PE (2000) The Protein Data Bank. Nucleic Acids Res 28(1):235–242
5. Yang JM, Tung CH (2006) Protein structure database search and evolutionary classification. Nucleic Acids Res 34(13):3646–3659
6. Yeh JS, Chen DY, Chen BY, Ouhyoung M (2005) A web-based three-dimensional protein retrieval system by matching visual similarity. Bioinformatics 21(13):3056–3057
7. Sael L, Li B, La D, Fang Y, Ramani K, Rustamov R, Kihara D (2008) Fast protein tertiary structure retrieval based on global surface shape similarity. Proteins 72(4): 1259–1273
8. Kihara D, Sael L, Chikhi R, Esquivel-Rodriguez J (2011) Molecular surface representation using 3D Zernike descriptors for protein shape comparison and docking. Curr Protein Pept Sci 12(6):520–530
9. Sael L, Kihara D (2012) Detecting local ligand-binding site similarity in nonhomologous proteins by surface patch comparison. Proteins 80(4):1177–1195
10. Chikhi R, Sael L, Kihara D (2011) Protein binding ligand prediction using moments-based methods. Protein function prediction for Omics Era, Springer, New York
11. Sael L, Kihara D (2010) Improved protein surface comparison and application to low-resolution protein structure data. BMC Bioinformatics 11 Suppl 11:S2
12. Venkatraman V, Chakravarthy PR, Kihara D (2009) Application of 3D Zernike descriptors to shape-based ligand similarity searching. J Cheminform 1:19
13. Venkatraman V, Yang YD, Sael L, Kihara D (2009) Protein-protein docking using region-based 3D Zernike descriptors. BMC Bioinformatics 10:407

14. Esquivel-Rodriguez J, Yang YD, Kihara D (2012) Multi-LZerD: multiple protein docking for asymmetric complexes. Proteins 80(7):1818–1833
15. La D, Esquivel-Rodriguez J, Venkatraman V, Li B, Sael L, Ueng S, Ahrendt S, Kihara D (2009) 3D-SURFER: software for high-throughput protein surface comparison and analysis. Bioinformatics 25(21):2843–2844
16. Li B, Turuvekere S, Agrawal M, La D, Ramani K, Kihara D (2008) Characterization of local geometry of protein surfaces with the visibility criterion. Proteins 71(2):670–683
17. Huang B, Schroeder M (2006) LIGSITE^{csc}: predicting ligand binding sites using the Connolly surface and degree of conservation. BMC Struct Biol 6:19
18. Orengo CA, Michie AD, Jones S, Jones DT, Swindells MB, Thornton JM (1997) CATH—a hierarchic classification of protein domain structures. Structure 5(8):1093–1108
19. Shindyalov IN, Bourne PE (1998) Protein structure alignment by incremental combinatorial extension (CE) of the optimal path. Protein Eng 11(9):739–747
20. Andreeva A, Howorth D, Chandonia JM, Brenner SE, Hubbard TJ, Chothia C, Murzin AG (2008) Data growth and its impact on the SCOP database: new developments. Nucleic Acids Res 36(Database issue):D419–D425
21. Connolly ML (1983) Solvent-accessible surfaces of proteins and nucleic acids. Science 221(4612):709–713
22. Novotni M, Klein R (2003) 3D Zernike descriptors for content based shape retrieval. Paper presented at the proceedings of the 8th ACM symposium on solid modeling and applications, Seattle
23. Connolly ML (1993) The molecular surface package. J Mol Graph 11(2):139–141
24. Barber CB, Dobkin DP, Huhdanpaa H (1996) The Quickhull algorithm for convex hulls. ACM T Math Software 22(4):469–483
25. Prashar V, Bihani S, Das A, Ferrer JL, Hosur M (2009) Catalytic water co-existing with a product peptide in the active site of HIV-1 protease revealed by X-ray structure analysis. PLoS One 4(11):e7860
26. Miller JF, Andrews CW, Brieger M, Furfine ES, Hale MR, Hanlon MH, Hazen RJ, Kaldor I, McLean EW, Reynolds D, Sammond DM, Spaltenstein A, Tung R, Turner EM, Xu RX, Sherrill RG (2006) Ultra-potent P1 modified arylsulfonamide HIV protease inhibitors: the discovery of GW0385. Bioorg Med Chem Lett 16(7):1788–1794

Chapter 9

SPOT-Seq-RNA: Predicting Protein–RNA Complex Structure and RNA-Binding Function by Fold Recognition and Binding Affinity Prediction

Yuedong Yang, Huiying Zhao, Jihua Wang, and Yaoqi Zhou

Abstract

RNA-binding proteins (RBPs) play key roles in RNA metabolism and post-transcriptional regulation. Computational methods have been developed separately for prediction of RBPs and RNA-binding residues by machine-learning techniques and prediction of protein–RNA complex structures by rigid or semiflexible structure-to-structure docking. Here, we describe a template-based technique called SPOT-Seq-RNA that integrates prediction of RBPs, RNA-binding residues, and protein–RNA complex structures into a single package. This integration is achieved by combining template-based structure-prediction software, SPARKS X, with binding affinity prediction software, DRNA. This tool yields reasonable sensitivity (46 %) and high precision (84 %) for an independent test set of 215 RBPs and 5,766 non-RBPs. SPOT-Seq-RNA is computationally efficient for genome-scale prediction of RBPs and protein–RNA complex structures. Its application to human genome study has revealed a similar sensitivity and ability to uncover hundreds of novel RBPs beyond simple homology. The online server and downloadable version of SPOT-Seq-RNA are available at http://sparks-lab.org/server/SPOT-Seq-RNA/.

Key words Fold recognition, Binding affinity, Protein–RNA complex structure, Template-based structure prediction, Knowledge-based energy function, Protein–RNA interactions, RNA-binding proteins, Torsion-angle prediction, Solvent accessible surface area, Prediction, SPOT-Seq-RNA

1 Introduction

The majority of the human genome is coded for RNA transcripts. Only tiny fractions of these RNA transcripts are messenger RNAs that code for proteins. All RNA transcripts, most with unknown functions, are regulated by RNA-binding proteins (RBPs) from birth (transcription) to death (degradation). Thus, locating all RBPs in a genome and determining protein–RNA complex structures are key steps for understanding the mechanism of post-transcriptional regulation and mapping the network of protein–RNA interactions.

Daisuke Kihara (ed.), *Protein Structure Prediction*, Methods in Molecular Biology, vol. 1137,
DOI 10.1007/978-1-4939-0366-5_9, © Springer Science+Business Media New York 2014

It is difficult to locate RBPs and determine their protein–RNA complex structures experimentally due to high flexibility of RNA structures and the difficulty associated with crystallization of complex structures. Despite this difficulty, there is a steady increase in the number of protein–RNA complex structures deposited in the protein data bank from 45 in 2001 to 180 in 2011 (nonredundant at 90 % sequence identity or less) [1]. Moreover, hundreds of novel, unconventional, or moonlighting RBPs have been discovered [2–4]. Experimental discovery of new RBPs and determination of protein–RNA complex structures, however, is costly and inefficient. There is a need for the development of highly accurate bioinformatics tools for predicting RNA-binding function and protein–RNA complex structures.

Most methods developed for predicting RBPs are based on machine-learning methods that employ information of protein sequences and/or known protein structures [5, 6]. Meanwhile, docking techniques for protein–RNA interactions have been developed by using a scoring/energy function for protein–RNA interaction [7–10]. Here, we describe SPOT-Seq-RNA, a template-based technique that combines predictions of protein–RNA complex structure and binding affinity [11]. More specifically, SPOT-Seq-RNA employs a template library of nonredundant protein–RNA complex structures and attempts to match the query sequence to the protein structures in protein–RNA complexes by the fold recognition technique SPARKS X [12]. Significant matches will be employed to predict the complex structures between a target sequence and template RNA as well as the binding affinity of the complex.

In SPOT-Seq-RNA, structure prediction is performed by the latest version of our fold recognition technique SPARKS X [12], which was among the best performing single automatic servers in several critical assessment of structure prediction (CASP) meetings (CASP 6 [13], CASP 7 [14], and CASP 9 [12]). SPARKS X is a multi-dimensional probabilistic matching between sequence profiles generated from PSI-BLAST [15] for query and template sequences and between structural features of a template and those predicted by SPINE X [16–18] for a query sequence. Predicted structural features include secondary structure [17], backbone torsion angles [16], and residue solvent accessibility [18]. For binding affinity prediction, we extracted a knowledge-based energy function, DRNA, from protein–RNA complex structures [19] based on a distance-scaled finite ideal-gas reference (DFIRE) state [20]. The DFIRE reference state was found to be one of the best reference states for deriving knowledge-based energy functions for folding and binding studies [21, 22]. While many template-based structure-prediction methods and knowledge-based energy functions for protein–RNA interactions exist, the coupling between fold recognition by SPARKS X and binding affinity prediction by DRNA in SPOT-Seq-RNA provides the first dedicated high-resolution function prediction for RBPs.

SPOT-Seq-RNA was cross-validated by leave-homology-out and independently tested by several datasets [11]. It was found to significantly improve over a sequence-to-profile search technique, PSI-BLAST [15], and a profile-to-profile search technique, HHPRED [23], in discriminating RBPs from non-RBPs. It was also shown to be far more sensitive and accurate in detecting RBPs than machine-learning-based techniques, while having similar accuracy to the best machine-learning techniques for RNA-binding site prediction [24]. More importantly, SPOT-Seq-RNA can provide a reasonably accurate prediction of protein–RNA complex structure (77 % predicted structures having root-mean-squared distance of 4 Å or less) [11]. More recently, SPOT-Seq-RNA was applied to the human genome and independently tested by mRNA-binding proteins from a proteomic experiment [25]. Discovery of more than 2,000 novel RBPs in the human genome and validation of the results in messenger-RBPs by the proteomic experiment [4] confirm the usefulness of SPOT-Seq-RNA in predicting novel RBPs beyond simple sequence homology and modeling of their complex structures.

2 Materials

2.1 Software

A software package is downloadable from our homepage with a shortcut link: http://sparks-lab.org/yueyang/download/index.php?Download=SPOT-Seq-RNA.tbz. This package as shown in Fig. 1 integrates one external program PSI-BLAST [15] (*see* **Note 1**)

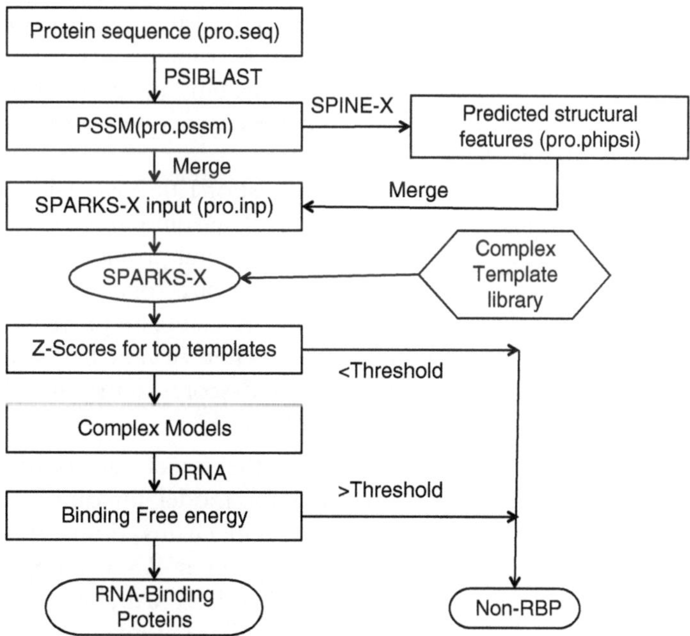

Fig. 1 The flow chart of SPOT-Seq-RNA

and three in-house-built programs: SPINE X (structural property prediction) [16, 17], SPARKS X (template-based structure prediction) [12], and DRNA (binding affinity prediction) [19].

1. An external program, PSI-BLAST, and protein NR database [15] are employed to generate a position-specific scoring matrix (PSSM) or sequence profile that is a required input for programs SPINE X and SPARKS X (*see* **Note 2** to skip this step if a PSSM file is pre-calculated for the query sequence).
2. An in-house-made program, SPINE X [16–18], is applied to predict the secondary structure, torsional angles (φ and ψ), and the solvent accessible surface area (ASA). SPINE X is a neural-network predictor that couples secondary structure prediction with predictions of solvent accessibility and backbone torsion angles in an iterative manner. SPINE X was tested with a dataset of 2,640 proteins and achieved an 82.0 % accuracy in secondary structure prediction based on tenfold cross validation. SPINE X can also be downloaded separately from our homepage.
3. SPARKS X is a template-based structure-prediction program. The program is employed to search for the best match between a query sequence and a template structure in the template database of protein–RNA complex structures. The statistically significant alignments from the best match (or matches) are utilized to construct complex structure models between the query and RNA of the template.
4. DRNA scoring function is used to calculate binding affinity. DRNA is a statistical energy function extracted from 174 protein–RNA complex structures with a DFIRE state [19]. It predicts the binding affinity based on the complex structure model between the query and template RNA.

2.2 Databases for RNA-Binding Proteins and Non-RNA-Binding Proteins

1. A prebuilt list of 1,052 RNA-binding domains and chains and 5,766 non-RNA-binding chains were prepared. The files for template structural profiles for both RBPs and non-RBPs are located in the directory "TPL_input." Here the database of RBPs contains template proteins in complex with their binding RNAs, while all non-RBPs serve as background statistics to calculate Z-scores to measure the significance of the matching template.
2. For RBPs the structural coordinates in PDB format were provided for model building. The coordinate files contain 1,052 protein chains/domains as well as 632 RNAs from respective protein–RNA complexes, stored in directories "domains" and "RNA0," respectively. Each protein file contains one chain or domain, while RNA coordinate files contain all RNA chains in the protein complex. These protein

coordinate files are employed as the templates to build the structure model of the query protein, while the RNA coordinates will be directly copied (with the same orientation as in the template complex structure) as the RNA conformation in the complex structure model for the query protein.

3 Methods

To describe the automated prediction pipeline as shown in Fig. 1, we used a protein, Bacillus halodurans RNase H catalytic domain mutant D132N (PDB id: chain B of 1zbi) as an example. This protein is RNase H and belongs to a nucleotidyl transferase superfamily, which includes transposase, retroviral integrase, Holliday junction resolvase, and RISC nuclease Argonaute [26].

3.1 Input and PSSM Generation

The only input required for SPOT-Seq-RNA is the query protein sequence in FASTA format (*see* **Note 3**). Figure 2a displays the input window for the web-based server that allows the cut-and-paste of the protein sequence RNase H. File upload is also allowed. Only one sequence per run is allowed for input. This sequence is subsequently passed to PSI-BLAST [15] to search for homologous sequences of the query sequence and to generate the position-specific substitution matrix (PSSM), which is constructed by three iterations of searching (*E* value less than 0.001) against the nonredundant (NR) sequence database.

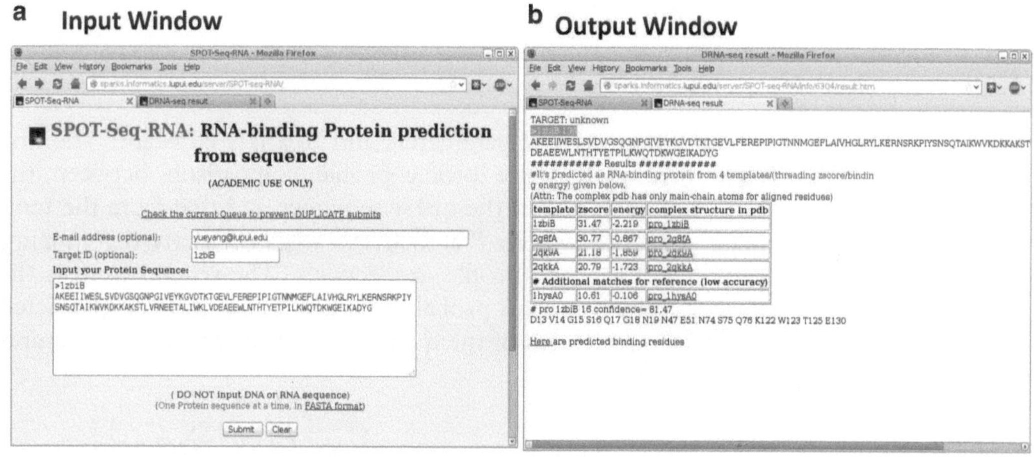

Fig. 2 The input and result windows of the SPOT-Seq-RNA server for the query protein Bacillus halodurans RNase H (PDBID: 1zbiB)

3.2 Structural Profile Preparation for SPARKS X

The PSSM file (either given or generated from PSI-BLAST) above is first employed by SPINE X to predict protein structural properties, including secondary structure (in three states), torsional angles (φ and ψ), and the ASA, along with their respective confidence scores. SPINE X is a neural-network predictor that utilizes a Perl script file to automatically call five separate predictors that were compiled by the Intel Fortran compiler. Structural properties predicted by SPINE X together with PSSM are stored in a profile file (pro.inp) as input for SPARKS X. In this profile file, the first line indicates the residue number (NRES) of the query protein. The second line contains the sequence in one-letter code. The next 20 lines are the inverse value of PSSM for 20 amino-acid residue types at all sequence positions (20 times NRES). These 20 lines are arranged alphabetically based on residue names (ACDE…Y). Here, the inverse value of PSSM is used to reduce the number of characters in the file as we noticed that most values in the PSSM are negative. After that, another 20 lines are the probability of 20 amino acids at each sequence position. Then, the next four lines are predicted probabilities of secondary structures (three states of coil, helix, and sheet, CHE), φ, ψ, and relative ASA. These structural properties are followed by predicted confidence scores for secondary structure, φ and ψ, respectively. In the current version, the confidence score for ASA is pre-calculated based on amino-acid residue types.

3.3 Scanning Over All Templates by SPARKS X

The above sequence and structural profiles for the query sequence are employed by SPARKS X to compare with corresponding profiles of all template structures in the template library (*see* Subheading 2.2). The raw profile–profile alignment score in SPARKS X is calculated as follows:

$$
\begin{aligned}
S(i,j) = & -\frac{1}{200}\left[F_{query}^{Seq}(j)\cdot M_{template}^{Seq}(i) + F_{template}^{Seq}(i)\cdot M_{query}^{Seq}(j)\right] \\
& + w_1 E\left(SS_t(i) \mid SS_q(j), C_{SS,q}(j)\right) \\
& + \sum_{k=2}^{4} w_k E\left(\Delta_{ij}^{k} \mid C_{k,q}(j)\right) + S_{shift}
\end{aligned}
\quad (1)
$$

where w_k are weight parameters and S_{shift} is a constant. The first term in Eq. 1 is the profile–profile comparison between the sequence profile from the query sequence and that from the template sequence, where F and M are sequence-derived frequency profile and log odd profile, respectively. The second term is the energy term based on probabilistic matching between predicted secondary structures of the query and actual secondary structures of the template:

$$
E\left(SS_t(i) \mid SS_q(j), C_{SS,q}(j)\right) = -\ln\left(\frac{P\left(SS_t \mid SS_q, C_{ss,q}\right)}{P\left(SS_t\right)}\right),
$$

where $P(SS_t|SS_q, C_{ss,q})$ is the probability of the predicted secondary structure SS_q by SPINE X with confidence score $C_{SS,q}$ for a native secondary structure SS_t. Similarly, the next three terms are the energy terms based on probabilistic matching between other structural properties:

$$E\left(\Delta^k \mid C_{k,q}\right) = -\ln\left(\frac{P\left(\Delta^k \mid C_{k,q}\right)}{P^0\left(\Delta^k \mid C_{k,q}\right)}\right),$$

where $P(\Delta k|C_{k,q})$ is the probability of the difference Δk between the predicted properties and corresponding native values with a confidence score of $C_{k,q}$. The reference probability $P0(\Delta k|C_{k,q})$ is obtained by comparing the predicted values to all native values in a dataset as described below. There are a total of three terms with $k=2$ for real-value φ value, $k=3$ for real-value ψ value, and $k=4$ for real-value solvent accessibility. All energy terms were obtained from a nonredundant data set of 2,479 proteins with length less than 500 amino acids from the original SPINE database [25 % sequence identity cutoff, X-ray resolution of 3 Å or higher, and no unknown structural regions] [27].

The raw alignment scores optimized by dynamic programming techniques for all templates are saved in the file called "pro.out," in which each line contains the template name, the raw alignment score, the total alignment length including gaps, the number of gaps in two termini, the start and end positions of the query chain segment with effective alignment, and the number of exactly matched residue types in the alignment. The number of gaps can be positive (gaps in the query protein) or negative (gaps in the template protein). The sequence position begins counting from zero.

3.4 Selecting Statistically Significant Matching Templates

From the alignment raw score, the *Z-score* was calculated based on a normalized score $S_{\text{norm}} = S_{\text{raw}}/L^{\alpha}$ using the standard definition: $Z\text{-}score = (S_{\text{norm}} - S_{\text{ave}})/\Delta S$, where S_{raw} and L are the raw alignment score and alignment length (i.e., the second and third column in the pro.out file); α is 0.75; and S_{ave} and ΔS are the average value and standard error of the normalized score on all templates. A higher Z-score indicates a highly significant matching template from the average templates. Based on our previous statistics, templates with Z-scores of six or higher have 90 % probability of having the same structural fold as the query protein.

By default, the program will record five or more templates with the highest Z-score or Z-scores greater than eight in file "pro.zs1." In the file, the first column is the calculated Z-score followed by the query protein name and raw alignment scores that are output by SPARKS X for each template. These templates will be subjected to model building and binding affinity evaluation.

3.5 Building and Evaluating Protein–RNA Complex Models

All top matching templates in the file "pro.zs1" are used to build complex models. The model structures are built based on the alignment between the query and template sequences (*see* **Note 4**). The coordinates of the main-chain and Cβ atoms (if present) of residues in the template will be copied to the corresponding aligned residues in the query protein. If the Cβ of a residue except GLY is missing in the template, the Cβ atom will be built based on the coordinates of the three main-chain heavy atoms (N, Cα, C). For those query residues not aligned to template residues, they will be ignored. The final protein model copies the RNA structure from the template to produce the complex structure model. All complex structure models are saved in separate files in PDB format (e.g., a complex built using template "2qk9A" will be saved in file "pro_2qk9A.pdb").

From these complex structure models, the binding free energy will be evaluated by using the program DRNA. The pairwise distance-dependent energy between an atom of an amino acid residue and an atom in an RNA base is

$$u_{ij}^{DRNA} = \begin{cases} -\eta \ln \dfrac{N_{obs}(i,j,r)}{\left(\dfrac{f_i^v(r) f_j^v(r)}{f_i^v(r_{cut}) f_j^v(r_{cut})}\right)^{\beta} \left(\dfrac{r}{r_{cut}}\right)^{\alpha} N_{obs}(i,j,r_{cut})}, r < r_{cut} \\ 0, r \geq r_{cut} \end{cases}$$

where $\alpha = 1.61$, $\beta = 0.5$, r_{cut} is the interaction cutoff distance (15 Å), and the volume-fraction factor $f_i^v(r) = \sum_j N_{obs}^{Protein\ RNA}(i,j,r) / \sum_j N_{obs}^{All}(i,j,r)$, $N_{\text{obs}}(i,j,r)$ is the number of observed pairs of atoms i and j at a given distance r from a database of protein–RNA complex structures. We employed residue/base-specific atom types with a total of 253 atom types (167 for protein and 86 for RNA). We also set the factor η arbitrarily to 0.01 to control the magnitude of the energy score. The statistics of $N_{\text{obs}}(i,j,r)$ is saved in the file "dfire_RNA." The binding free energy of a complex structure model is obtained by summing the interactions between any RNA atoms and protein atoms of main-chain atoms and Cβ only with a distance less than 6.0 Å (**Note 5**). The calculated binding free energy together with the Z-score for all complexes is recorded in the file "pro.zs_en."

3.6 Detecting RNA-Binding Proteins

The query protein is an RBP when both Z-score and energy thresholds are satisfied for at least one complex structure model. The final output file, "pro.result," contains the template name, Z-score and the estimated binding free energy of the protein–RNA complex structure. The complex structure is then employed to predict residues that interact with RNA (binding residue prediction). The binding residues are defined if any atom of the residue

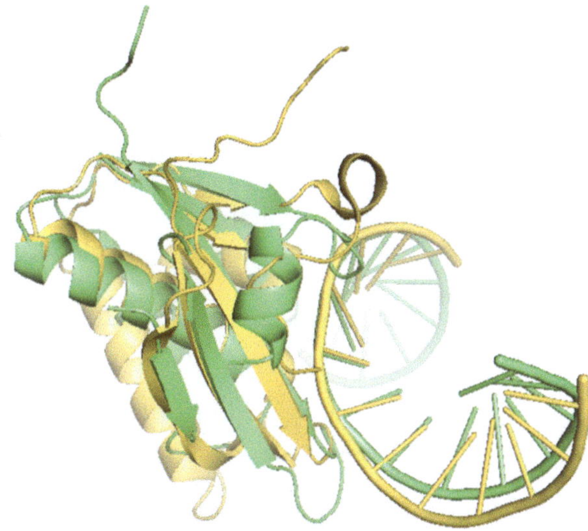

Fig. 3 The predicted model based on template 2qk9A for protein Bacillus halodurans RNase H (*colored in green*) is structurally aligned by SPalign to the native structure (*colored in yellow*). One complementary DNA chain has been removed

is less than 4.5 Å to any RNA atoms. For a balance of coverage and accuracy of the prediction, we have set a threshold of 8.04 and −0.565 for Z-score and binding free energy, respectively. These thresholds were obtained in our benchmark studies by maximizing the Matthews correlation coefficient (MCC) for two-state prediction of RBPs [11].

If not a single template is found to satisfy both thresholds, the query protein will be considered as a non-RBP. However, we continue to present the top five matched templates and predicted complex structure models because low sensitivity (about 46 %) may incorrectly predict some RBPs as non-RBPs despite correct prediction of complex structures. Users may have additional biological information to judge correctness of the complex structure model and function prediction (*see* **Note 6**).

3.7 SPOT-Seq-RNA Input/Output

Figure 2 shows the input and output windows of the SPOT-Seq-RNA server at http://sparks-lab.org/server/SPOT-Seq-RNA/. This output is based on the query protein Bacillus halodurans RNase H (For running time, *see* **Note 7**). There are four matching templates within both Z-score and binding thresholds. Thus, this protein is predicted as an RBP. Predicted complex structural models are listed according to respective templates. After filtering homologous templates, 2qk9A (17.6 % sequence identity to the query sequence) was selected to demonstrate the overall accuracy of prediction. Figure 3 displays the structurally aligned predicted and native complex structures by SPalign [28]. One hundred and seven residues out of 136 residues (79 %) in the query protein are aligned with its actual native structure with RMSD 2.8 Å. In addition to the four templates within the thresholds, there is a template "1hysA0" that satisfied the Z-score but not the binding threshold. This illustrates the possibility of false negatives despite accurate structure prediction (*see* **Note 6**). Results from the online server can be delivered to an email address (**Note 8**).

4 Notes

1. The recent blast package can be downloaded from: ftp://ftp.ncbi.nlm.nih.gov/blast/executables/release/LATEST/

   Select the appropriate executable version for your system (e.g., blast-2.2.26-x64-linux.tar.gz for 64 bit Linux). The NR database can be downloaded from ftp://ftp.ncbi.nih.gov/blast/db/nr.XX.tar.gz. As of Feb 23, 2013, XX includes ten numbers from 00 to 09, and each file is about 700 megabytes.
2. The package requires PSSM from PSI-BLAST. If a pre-calculated PSSM was prepared, PSI-BLAST can be skipped to save time. The user can choose to input a pre-calculated PSSM with option "-pssm" for the locally installed version.
3. The query sequence must be a protein sequence in FASTA format. The gene in the DNA/RNA sequence has to be converted to amino acids sequence first. Unknown amino acids (e.g., X) must be removed.
4. SPARKS X has the option "-print level" to control different levels of outputs. The default level (0) only outputs the alignment score and length information. This will reduce the size of the output file in the template scanning step. A level equal to or greater than two will also print out the alignment between query and templates.
5. Here unaligned residues and the side-chain atoms except Cβ atoms are excluded for interaction calculations so that we can prevent large fluctuation in predicted binding affinities due to possible atomic clashes between RNA and modeled side-chains or modeled missing residues. A new version is in progress to relax modeled side-chain and missing residues so that we can estimate the protein–RNA-binding affinity based on all interactions between protein and RNA molecules.
6. Some predicted non-RBPs within the boundary of thresholds may be false negatives and have correctly predicted binding models. The strict cutoffs in Z-score and binding affinity were determined to maximize the MCC in our benchmark (low sensitivity around 46 % but high precision at 84 %) [11]. For those templates with a Z-score greater than six but less than eight, the model protein structure is likely correct.

7. The running time depends on the size of the query protein. For the example given here (1zbiB, 136 residues), it takes 22 min on an Intel Pentium 4 3.4GHz, in which 14 min are due to PSI-BLAST.

8. For online service, the results can be obtained from the webpage directly or from email if an email address is given. To save computing resources, please do not submit query sequences more than once. The status of your job can be found by clicking the link "Check the current Queue to prevent DUPLICATE submits" on the main webpage. The result of your job will only be kept for 1 month after completion.

Acknowledgments

Funding for this work was supported by the National Institutes of Health grants [GM R01 085003 and GM R01 067168 (Co-PI) to Y.Z.] and by the National Natural Science Foundation of China [grant 61271378 to J.W.].

References

1. Bernstein FC, Koetzle TF, Williams GJ, Meyer EF Jr, Brice MD, Rodgers JR, Kennard O, Shimanouchi T, Tasumi M (1977) The Protein Data Bank: a computer-based archival file for macromolecular structures. J Mol Biol 112: 535–542
2. Tsvetanova NG, Klass DM, Salzman J, Brown PO (2010) Proteome-wide search reveals unexpected RNA-binding proteins in Saccharomyces cerevisiae. PLoS One 5:e12671
3. Scherrer T, Mittal N, Janga SC, Gerber AP (2010) A screen for RNA-binding proteins in yeast indicates dual functions for many enzymes. PLoS One 5:e15499
4. Castello A, Fischer B, Eichelbaum K, Horos R, Beckmann BM, Strein C, Davey NE, Humphreys DT, Preiss T, Steinmetz LM et al (2012) Insights into RNA biology from an Atlas of mammalian mRNA-binding proteins. Cell 149:1393–1406
5. Puton T, Kozlowski L, Tuszynska I, Rother K, Bujnicki JM (2012) Computational methods for prediction of protein-RNA interactions. J Struct Biol 179(3):261–8
6. Walia RR, Caragea C, Lewis BA, Towfic FG, Terribilini M, El-Manzalawy Y, Dobbs D, Honavar V (2012) Protein-RNA interface residue prediction using machine learning: an assessment of the state of the art. BMC Bioinformatics 13:89
7. Perez-Cano L, Solernou A, Pons C, Fernandez-Recio J (2010) Structural prediction of protein-RNA interaction by computational docking with propensity-based statistical potentials. Pac Symp Biocomput 15:269–280
8. Zheng S, Robertson TA, Varani G (2007) A knowledge-based potential function predicts the specificity and relative binding energy of RNA-binding proteins. FEBS J 274: 6378–6391
9. Tuszynska I, Bujnicki JM (2011) DARS-RNP and QUASI-RNP: new statistical potentials for protein-RNA docking. BMC Bioinformatics 12:348
10. Setny P, Zacharias M (2011) A coarse-grained force field for Protein-RNA docking. Nucleic Acids Res 39:9118–9129
11. Zhao H, Yang Y, Zhou Y (2011) Highly accurate and high-resolution function prediction of RNA binding proteins by fold recognition and binding affinity prediction. RNA Biol 8: 988–996
12. Yang Y, Faraggi E, Zhao H, Zhou Y (2011) Improving protein fold recognition and template-based modeling by employing probabilistic-based matching between predicted

one-dimensional structural properties of the query and corresponding native properties of templates. Bioinformatics 27:2076–2082

13. Zhou HY, Zhou Y (2005) SPARKS 2 and SP3 servers in CASP 6. Proteins 61:152–156
14. Liu S, Zhang C, Liang SD, Zhou Y (2007) Fold recognition by concurrent use of solvent accessibility and residue depth. Proteins 68: 636–645
15. Altschul SF, Madden TL, Schaffer AA, Zhang JH, Zhang Z, Miller W, Lipman DJ (1997) Gapped BLAST and PSI-BLAST: a new generation of protein database search programs. Nucleic Acids Res 25:3389–3402
16. Faraggi E, Yang YD, Zhang SS, Zhou Y (2009) Predicting continuous local structure and the effect of its substitution for secondary structure in fragment-free protein structure prediction. Structure 17:1515–1527
17. Faraggi E, Zhang T, Yang Y, Kurgan L, Zhou Y (2011) SPINE X: improving protein secondary structure prediction by multi-step learning coupled with prediction of solvent accessible surface area and backbone torsion angles. J Comput Chem 33:259–263
18. Faraggi E, Xue B, Zhou Y (2009) Improving the prediction accuracy of residue solvent accessibility and real-value backbone torsion angles of proteins by guided-learning through a two-layer neural network. Proteins 74: 847–856
19. Zhao HY, Yang YD, Zhou YQ (2011) Structure-based prediction of RNA-binding domains and RNA-binding sites and application to structural genomics targets. Nucleic Acids Res 39:3017–3025
20. Zhou HY, Zhou Y (2002) Distance-scaled, finite ideal-gas reference state improves structure-derived potentials of mean force for structure selection and stability prediction. Protein Sci 11:2714–2726
21. Zhou Y, Zhou HY, Zhang C, Liu S (2006) What is a desirable statistical energy function for proteins and how can it be obtained? Cell Biochem Biophys 46:165–174
22. Zhou YQ, Duan Y, Yang YD, Faraggi E, Lei HX (2011) Trends in template/fragment-free protein structure prediction. Theor Chem Acc 128:3–16
23. Soding J, Biegert A, Lupas AN (2005) The HHpred interactive server for protein homology detection and structure prediction. Nucleic Acids Res 33:W244–W248
24. Zhao H, Yang Y, Zhou Y (2013) Prediction of RNA binding proteins comes of age from low resolution to high resolution. Mol Biosyst 9(10):2417–25
25. Zhao H, Yang Y, Janga SC, Kao C, Zhou Y (2013) Prediction and validation of the unexplored RNA-binding protein atlas of the human genome. Proteins, in press (doi: 10.1002/prot.24441)
26. Nowotny M, Gaidamakov SA, Crouch RJ, Yang W (2005) Crystal structures of RNase H bound to an RNA/DNA hybrid: substrate specificity and metal-dependent catalysis. Cell 121:1005–1016
27. Dor O, Zhou Y (2007) Achieving 80 % ten-fold cross-validated accuracy for secondary structure prediction by large-scale training. Proteins 66:838–845
28. Yang Y, Zhan J, Zhao H, Zhou Y (2012) A new size-independent score for pairwise protein structure alignment and its application to structure classification and nucleic-acid binding prediction. Proteins 80:2080–2088

Chapter 10

POODLE: Tools Predicting Intrinsically Disordered Regions of Amino Acid Sequence

Kana Shimizu

Abstract

Protein intrinsic disorder, a widespread phenomenon characterized by a lack of stable three-dimensional structure, is thought to play an important role in protein function. In the last decade, dozens of computational methods for predicting intrinsic disorder from amino acid sequences have been developed. They are widely used by structural biologists not only for analyzing the biological function of intrinsic disorder but also for finding flexible regions that possibly hinder successful crystallization of the full-length protein. In this chapter, I introduce *Prediction Of Order and Disorder by machine LEarning* (POODLE), which is a series of programs accurately predicting intrinsic disorder. After giving the theoretical background for predicting intrinsic disorder, I give a detailed guide to using POODLE. I then also briefly introduce a case study where using POODLE for functional analyses of protein disorder led to a novel biological findings.

Key words Intrinsically disordered protein, Prediction, Amino acid sequence, Machine learning, Position-specific scoring matrix, Amino acid composition

1 Introduction

1.1 Predicting Protein Intrinsic Disorder: Background and Purposes

It has been widely accepted that the function of a protein is determined by its three-dimensional (3D) structure, but this classical structure–function paradigm has been challenged by recent studies finding that many regions of proteins lack stable 3D structures under physiological conditions and that unfolded regions play important roles in various essential biological functions, such as cell signaling and transcription and translation [1–3]. These protein regions are called intrinsically disordered regions (IDRs), and proteins with the IDRs are thought to exert their function by using conformational flexibility [4, 5]. Since the binding mechanism of proteins with IDRs is very different from that of rigid proteins, IDRs are one of the most spotlighted topics in structural biology, and dozens of computational methods for predicting them from amino acid sequences have been developed [6, 7].

Daisuke Kihara (ed.), *Protein Structure Prediction*, Methods in Molecular Biology, vol. 1137, DOI 10.1007/978-1-4939-0366-5_10, © Springer Science+Business Media New York 2014

One purpose for predicting IDRs is to assist experiments, and IDR prediction has been used to guide many experiments revealing the functions of newly identified IDRs [8–13]. Prediction of IDRs also helps X-ray structure determination. IDRs often hinder successful crystallization of a full-length protein, which leads to failure in determining its structure. Crystallization of a protein with IDRs often becomes successful by predicting IDRs and eliminating them. Since IDRs often appear in domain linkers of protein domains [1, 14], IDR prediction helps to find domain boundaries of multi-domain proteins [15, 16]. The identification of the domain boundaries is an important first step for both experimental studies and protein 3D structure predictions especially for eukaryotic proteins, most of which contain multiple domains.

Predicting IDRs in large-scale database analyses is especially useful because the number of IDRs found experimentally is not large enough to reflect the genome-wide predisposition of IDRs. Several genome-wide analyses by IDR prediction have revealed that the frequency of proteins that include long IDRs is significantly greater in eukaryotic proteomes than it is in prokaryotic proteomes [14, 17–19]. These analyses also have suggested that IDRs are more prevalent in higher organisms that require more complex signaling and regulatory events. Another genome-wide study found that human transcription factors contain a high fraction of IDRs [20]. Since IDRs are thought to mediate protein–protein interactions (PPIs), analyzing the results of IDR prediction in combination with PPI network yielded many interesting biological insights. For example, previous studies have found that hub proteins are likely to contain IDRs and that the resultant conformational flexibility is the basis of their interaction with different partners according to the environmental conditions [21–24].

1.2 Computational IDR Prediction

The first computational analysis of IDRs was done by Dunker and Uversky and their coworkers, who reported that the ratios of hydrophilic residues and charged residues in unstructured proteins are significantly higher than those in structured proteins [25]. Similar analyses by several other groups also found that the amino acid sequences of IDRs have features different from those of the amino acid sequences of structured protein. Most IDR prediction tools have been based on this theoretical background [18, 26–32]. The input to an IDR prediction tool is an amino acid sequence and the output is a sequence of binary labels, each showing whether or not the corresponding amino acid is predicted to be included in an IDR. Most tools also assign a probability or score that indicates the confidence level for each prediction.

The main approach to predicting IDRs is an ab initio approach predicting the likelihood of intrinsic disorder from only the input amino acid sequence. The basic strategy of this approach is to identify features frequently appearing in the amino acid sequences of known IDRs and to predict that a region is intrinsically disordered if the input sequence for that region has those features. For example, amino acid compositions of IDRs are largely different from those of structured regions. If the amino acid composition of a protein sequence is more similar to that of IDRs than to that of structured regions, the sequence is considered to be disordered. Most of the existing programs implement this strategy by using machine learning techniques, such as support vector machines and neural networks, that efficiently combine multiple features to make accurate predictions. A typical IDR prediction method based on a machine learning technique consists of training and prediction stages.

1.2.1 Training Stage

In this stage a classifier that classifies a subsequence of an input sequence into an IDR or a structured region is constructed. Amino acid sequences whose IDRs and structured regions have been confirmed experimentally are collected. For each, a sliding-window moves from the C-terminus of the sequence toward the N-terminus until the window reaches the N-terminus of the sequence. At each position of the sliding-window, the subsequence within the window is extracted and assigned a label showing if it was extracted from an IDR or a structured region. The set of these labeled subsequences is called the training dataset. Each labeled subsequence is represented as a vector of predetermined features such as amino acid compositions, and the similarity of two subsequences is calculated by measuring the similarity of the corresponding vectors. The machine learning algorithm generates an optimal classifier by trying to reduce miss-classification in the given training dataset.

1.2.2 Prediction Stage

In this stage each residue of an input sequence is predicted to be included in an IDR or not. Subsequences of a given protein sequence are extracted by using sliding-window and each subsequence is represented as a feature vector in the same manner as in the training stage. Each subsequence is classified into an IDR or a structured region by using the classifier generated in the training stage. Per-residue predictions are based on the results of the classifications of the subsequences. For example, each residue is predicted to be included in an IDR if the number of disordered subsequences covering the residue is larger than that of structured subsequences covering the residue. An overview of the method is shown in Fig. 1.

Some of the IDR prediction tools based on the ab initio approach are POODLE-S [30, 32], POODLE-L [31], POODLE-W [18], PONDR VSL2 [33], DISOPRED [34], IUPred [35], RONN [36], and PrDOS [37].

Another popular approach to predicting IDRs is called the meta-approach. In this approach a prediction tool does not directly predict IDRs from the input sequence but instead runs several IDR prediction programs on the input sequence and makes a final

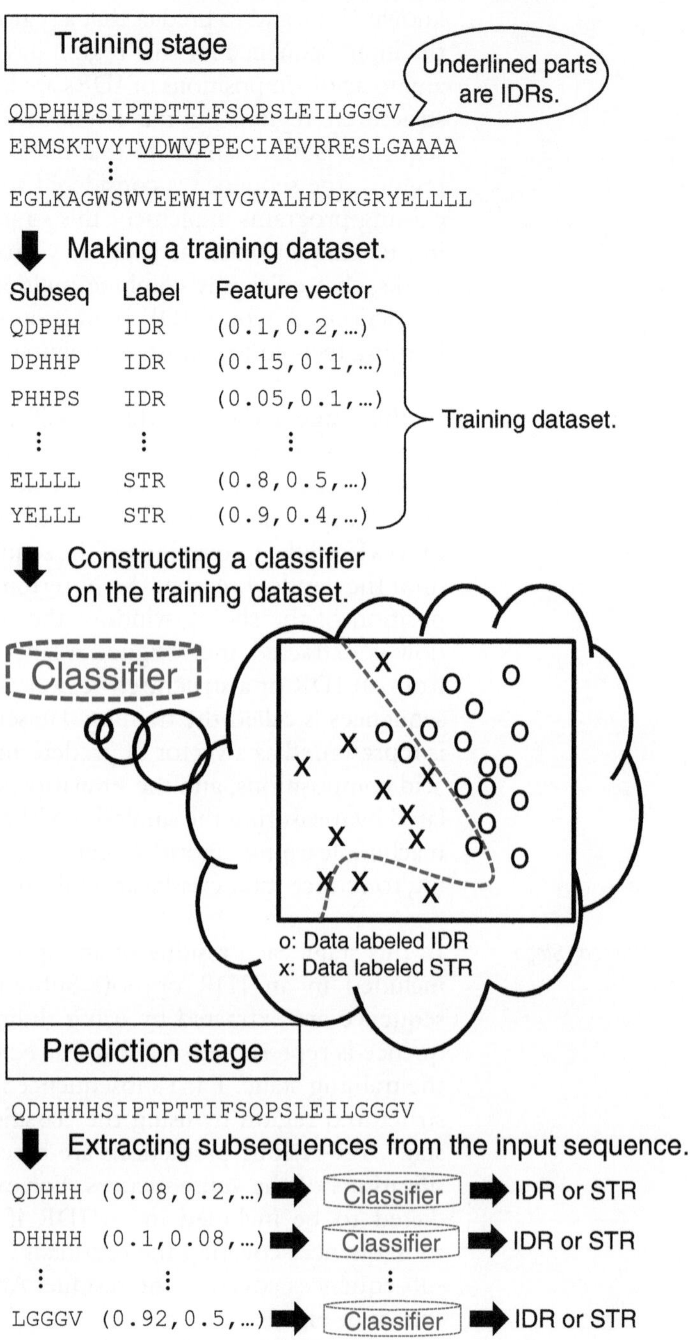

Fig. 1 Overview of a typical ab initio predictor based on machine learning

prediction by taking into account all of the results reported by those programs. Although a theoretical advantage for the meta-approach has not been shown, prediction tools based on it frequently give better results than ab initio prediction tools. Some of the IDR prediction tools based on the meta-approach are POODLE-I [38], PONDR-FIT [39], metaPrDOS [40], and metaDisorder [41].

1.3 Evaluation of IDR Prediction Tools

The ability of IDR prediction tools to make correct predictions is frequently evaluated by using the following measures:

$$\text{Sensitivity} = \frac{\text{TP}}{\text{TP} + \text{FN}},$$

$$\text{Specificity} = \frac{\text{TN}}{\text{TN} + \text{FP}},$$

$$\text{MCC} = \frac{\text{TP} \times \text{TN} - \text{FP} \times \text{FN}}{\sqrt{(\text{TP} + \text{FP})(\text{TP} + \text{FN})(\text{TN} + \text{FP})(\text{TN} + \text{FN})}},$$

where TP (true positives) is the number of disordered residues that are predicted to be disordered, TN (true negatives) is the number of structured residues that are predicted to be structured, FP (false positives) is the number of structured residues that are predicted to be disordered, and FN (false negatives) the number of disordered residues that are predicted to be structured. Tools with higher sensitivity are better able to detect disordered residues, and tools with higher specificity produce fewer false positives. There is a trade-off between sensitivity and specificity, and the MCC balances these two measures. The trade-off between sensitivity and specificity is represented by an ROC curve, which is a graph plotting sensitivity versus (1 – specificity). Area under an ROC curve (AUC) is one of the most effective measures. The most reliable evaluation of IDR prediction tools is the biennial experiment called Critical Assessment of Structure Prediction (CASP), which is a well-known blind test for protein structure prediction. In past CASPs, many IDR prediction methods were evaluated with several measures, including sensitivity, specificity, MCC, and AUC. Details of the evaluations are reported in journal papers [42–44].

1.4 POODLE

POODLE is a series of IDR prediction programs. As with other IDR prediction tools, the input is an amino acid sequence and the output is a sequence of binary labels and confidence scores. Each label shows whether or not the corresponding amino acid is predicted to be in an IDR, and each confidence score is the confidence level for that prediction.

Three of the four programs in the POODLE series—POODLE-S [30, 32], POODLE-L [31], and POODLE-W [18]—are based on the ab initio approach. As mentioned in the previous subsection, those programs use machine learning techniques. Two factors greatly affect the prediction accuracy when machine learning techniques are used. The first is the selection of the features that become the basis for discrimination. The second is the preparation of the training dataset. Since IDRs play various functional roles, the patterns appearing in amino acid sequences are also thought to vary depending on the functions. Therefore, the ideal ab initio predictor would identify the common features from separate datasets each of which share the same protein function and would be trained on those datasets.

In making the POODLE series, we separated training datasets into three types according to the length of the IDR, which is thought to be related to the function of the IDR. We also separately selected features in order to make each program fit to each type of training dataset. For example, POODLE-S uses scores derived from a position-specific scoring matrix (PSSM) in order to find short IDRs, while POODLE-L and POODLE-W use amino acid compositions for finding long IDRs and mostly disordered proteins. Both POODLE-S and POODLE-L use support vector machines, which are efficient supervised learning methods. For POODLE-W, a semi-supervised learning method that utilizes unlabeled training data to increase prediction accuracy is used because the amount of labeled training data (i.e., the number of proteins confirmed experimentally to be mostly disordered) is small. The programs in the POODLE series were carefully designed to cover all types of IDRs ranging from a single missing residue to a fully disordered protein. Details about each program are given in Subheading 4.

POODLE-I [38] is based on the meta-approach. It is a workflow system to predict IDRs from the results of other POODLE programs. One of the advantages of POODLE-I over other tools using the meta-approach is that all the programs that are used as sub-modules in POODLE-I are in the same server and the method POODLE-I uses was designed taking into account the detailed algorithm of each sub-module program. Many of other tools, in contrast, use prediction servers designed and maintained by different research groups. Prediction tools using the meta-approach potentially increase accuracy by taking into account the results of several types of ab initio methods, but their accuracy can decrease if any of the sub-modules cannot be used. POODLE-I is therefore much more stable than those other tools.

The POODLE series was evaluated in past CASPs, has been ranked as one of the top predictors [42–44].

The POODLE series has been used for analyzing many biological functions. For example, it was used to help design an experiment that found the proline-rich (PR) domain of Gab1 to be intrinsically disordered [12]. POODLE was also used for structural analysis of several proteins, including human papillomavirus proteins [45], ALK1 [46], BRMS1 [9], and human b-Gal [47]. Two prediction servers based on the meta-approach [40, 41] use POODLE-S as a sub-module. POODLE-S is used for predicting domain boundaries in a protein-structure-prediction-pipeline [15].

2 Materials

POODLE programs are provided by a web server accessible at http://mbs.cbrc.jp/poodle/poodle.html. The input to the server is an amino acid sequence written in standard single-letter code.

The server accepts both plain text and FASTA format. Predictions by POODLE-S, POODLE-W, or POODLE-I require the submission of an e-mail address to which the result is sent.

3 Methods

3.1 Running POODLE

Figure 2 shows a screen shot of the web page you see when you use a web browser to access the POODLE server (http://mbs.cbrc.jp/poodle/poodle.html). Only three steps are required for running POODLE.

Step 1: Selecting a program—As described in Subheading 1, POODLE provides several programs according to the purpose of the prediction. Select the one that fits your purpose. The features of each program are summarized in Subheading 4.

Step 2: Inputting an amino acid sequence—Prepare an amino acid sequence in plain text or in FASTA format, and input the sequence in the top form of the web page. If you have chosen POODLE-W in the first step, up to 50 sequences can be submitted as a single query by using the multiple FASTA format.

Step 3: Sending a query—Except when POODLE-L is chosen in **step 1**, prediction results are sent by e-mail. So for the other POODLE tools you must input, in the bottom form, the e-mail address to which you want your result sent. After inputting all the required information, click the "submit" button to send your query.

3.2 Presentation of Results

The result formats differ slightly for the different POODLE programs.

3.2.1 POODLE-S and POODLE-I

The results are sent by e-mail. The e-mail includes the URL of the graphical result, which is stored in the server and kept for 2 weeks. The *X*-axis in the result graph shows the position of the amino acids in the input sequence: the left-most position is the N-terminus of the sequence and the right-most position is the C-terminus of the sequence. The *Y*-axis shows, for each amino acid, the probability of being in a disordered region. The graph is interactive and it shows the probability and amino acid (in single-letter code) when the mouse pointer is overlapped on the graph line. An example result graph is shown in Fig. 3.

The e-mail also includes the result in text form. The lines between "METHOD ------ " and "END" show prediction results. The number of lines is equal to the length of the query amino acid sequence. The uppermost result line shows the result for the first amino acid in the submitted sequence, and the following lines consecutively show the results for the following amino acids in that sequence. In each line the first letter is the one-letter code for the

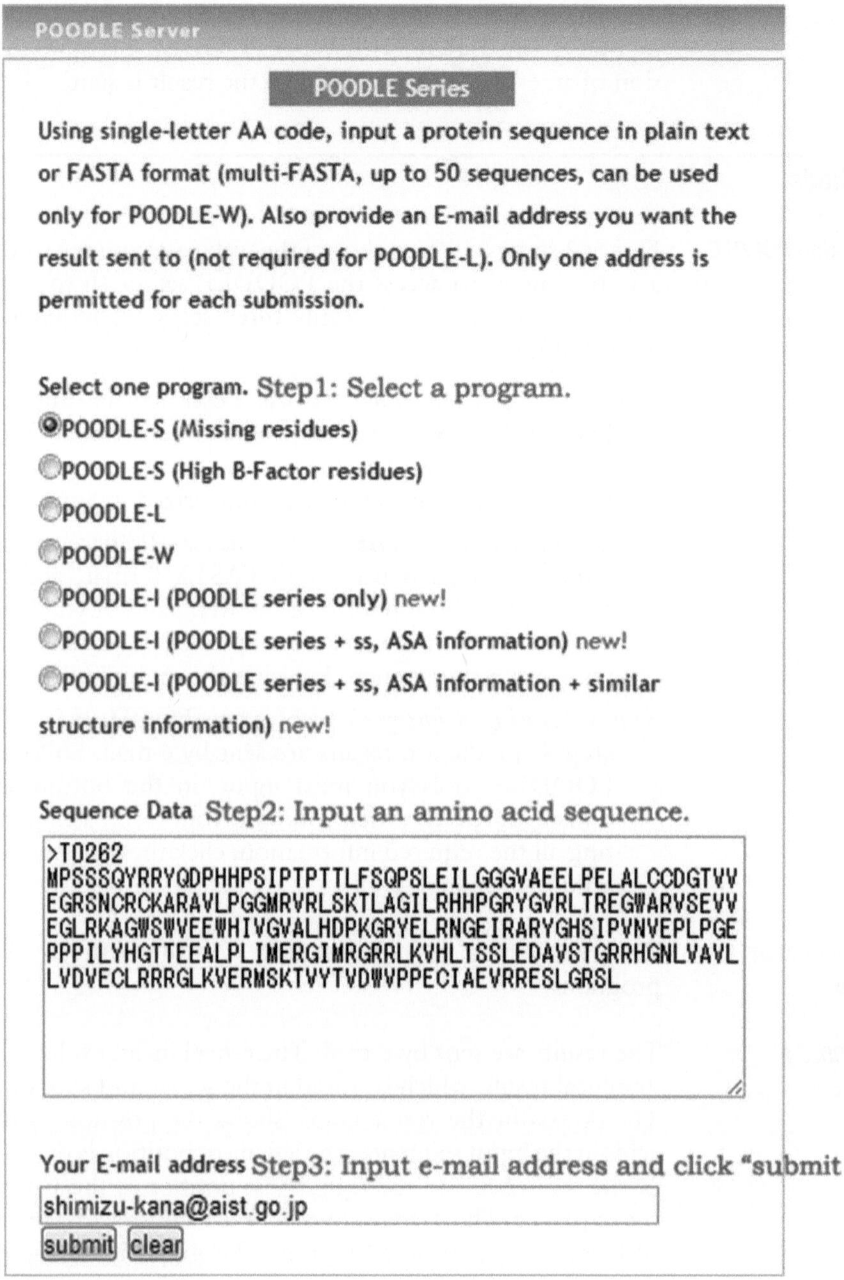

POODLE Server

POODLE Series

Using single-letter AA code, input a protein sequence in plain text or FASTA format (multi-FASTA, up to 50 sequences, can be used only for POODLE-W). Also provide an E-mail address you want the result sent to (not required for POODLE-L). Only one address is permitted for each submission.

Select one program. Step1: Select a program.

- POODLE-S (Missing residues)
- POODLE-S (High B-Factor residues)
- POODLE-L
- POODLE-W
- POODLE-I (POODLE series only) new!
- POODLE-I (POODLE series + ss, ASA information) new!
- POODLE-I (POODLE series + ss, ASA information + similar structure information) new!

Sequence Data Step2: Input an amino acid sequence.

```
>T0262
MPSSSQYRRYQDPHHPSIPTPTTLFSQPSLEILGGGVAEELPELALCCDGTVV
EGRSNCRCKARAVLPGGMRVRLSKTLAGILRHHPGRYGVRLTREGWARVSEVV
EGLRKAGWSWVEEWHIVGVALHDPKGRYELRNGEIRARYGHSIPVNVEPLPGE
PPPILYHGTTEEALPLIMERGIMRGRRLKVHLTSSLEDAVSTGRRHGNLVAVL
LVDVECLRRRGLKVERMSKTVYTVDWVPPECIAEVRRESLGRSL
```

Your E-mail address Step3: Input e-mail address and click "submit

shimizu-kana@aist.go.jp

submit clear

Fig. 2 Screen shot of input form

amino acid, the second letter shows whether the amino acid is in an ordered region or disordered region, and the number is the probability of its being in a disordered region. If this probability is more than 0.5, the second letter is "D"; otherwise it is "O." An example result is shown in Fig. 4.

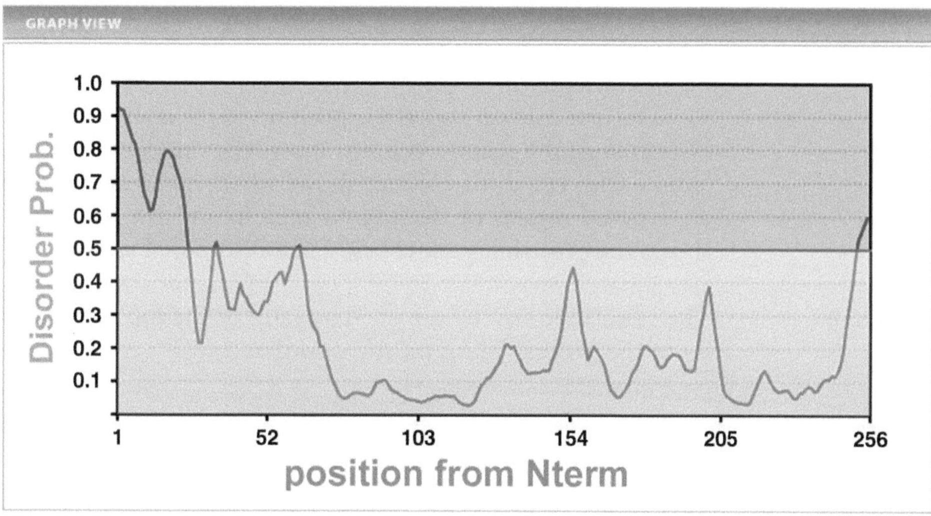

Fig. 3 Screen shot of graphical output

Your fasta Header: T0262

Graphical Output is accessible from here: URL to graphical output ↓
http://mbs.cbrc.jp/poodle/disorder/show.php?filename=2013_01_17_12_21_278284&email=shimizu-kana@aist.go.jp

```
PFRMAT DR
REMARK K. Shimizu, Y. Muraoka, S. Hirose, and T. Noguchi
REMARK "Feature Selection Based on Physicochemical Properties of
REMARK Redefined N-term Region and C-term Regions for Predicting Disorder"
REMARK Proc. of IEEE CIBCB 2005, pp262-267.
METHOD Prediction for short disorder using modified PSSM
METHOD ------
M D 0.928
P D 0.919    ← Amino acid, Prediction (D or O), Probability
S D 0.917
S D 0.89
S D 0.858
Q D 0.831
Y D 0.813
R D 0.773
R D 0.72
Y D 0.669
Q D 0.643
D D 0.613
P D 0.62
```

Fig. 4 Output of POODLE-S and POODLE-I

3.2.2 POODLE-L

The result itself is shown on the web page. The top figure is the interactive graph and the table in the middle of the page shows the detailed results.

The first column shows the position of the amino acid from the N-terminus, the second column shows the one-letter code for the amino acid, the third column shows whether the amino acid is in an ordered region or disordered region, and the fourth column shows the probability of its being in a disordered region.

3.2.3 POODLE-W

The results are sent by e-mail. The value at the top is the probability of the protein being mostly disordered.

4 Notes

As described in Subheading 3, the choice of program is important. The following program descriptions are summarized in Table 1.

POODLE-S: The main focus of POODLE-S is to predict the narrow trend of intrinsic disorder. It is designed to find IDRs more than a few amino acids long. It makes a multiple-sequence alignment with known proteins and a PSSM made from the alignments is used for prediction. It extracts related scores from the PSSM by using a sliding-window and uses those scores to determine whether or not the middle amino acid in each window is predicted to be in a disordered region. POODLE-S thus uses information derived from amino acid sequences similar to the input sequence. Since terminus regions of an amino acid sequence tend to be flexible, and these regions often include short-disordered regions, the sliding-window of five amino acids long is used for the terminus regions, and that of 15 amino acids long is used for the other internal region.

Table 1
Summary of POODLE series

	Main focus	Options
POODLE-S	Short trend of IDR	1. Find missing residues 2. Find residues whose B-factor is high
POODLE-L	Long trend of IDR	
POODLE-W	Protein-wide trend of IDR	
POODLE-I	General purpose (meta-approach)	1. Combine POODLE-S, -L, and -W 2. Combine POODLE-S, -L, -W, secondary structure prediction, solvent accessibility prediction, and coiled coil prediction 3. Combine POODLE-S, -L, -W, secondary structure prediction, solvent accessibility prediction, coiled coil prediction, and tertiary structure prediction

POODLE-S has two options. The first option offers to find "missing residues." For this option, POODLE-S is trained on the dataset where the missing residues of known protein structures are labeled as disordered and the other residues are labeled as structured. The second option offers to find residues whose B-factors are high compared with those of the other residues in the same sequence. For this option, POODLE-S is trained on a dataset where residues whose B-factors are high are labeled as disordered and the others are labeled as structured. Although these two options give similar results, the second (high-B-factor predictor) is expected to be more sensitive to amino acids with flexible conformations.

POODLE-L: The main focus of POODLE-L is to predict a broad trend of intrinsic disorder. It is designed to find IDRs whose length is more than 30 amino acids. It uses only amino acid-composition-related features obtained from an input sequence and does not rely on multiple-sequence-alignment information. POODLE-L also uses a sliding-window when it extracts features and makes a prediction. The size of the sliding-window is 40 amino acids.

POODLE-W: POODLE-W is used to find a protein-wide trend of intrinsic disorder. It determines whether or not the input sequence is mostly unstructured. It therefore gives only one score for each prediction. As mentioned in Subheading 1, it takes into account of unlabeled training data when it makes a prediction. Amino acid sequences included in Swiss-Prot are used as unlabeled training data.

POODLE-I: POODLE-I is a general-purpose IDR predictor combining several programs. It has three options. The first combines only the POODLE series programs. The second option combines those programs; the secondary structure prediction tools PSIPRED [48], JNET [49], and SABLE [50]; the solvent accessibility prediction tools JNET and SABLE; and the coiled coil prediction tool COILS [51]. The third option combines the POODLE series programs, secondary structure prediction tools, solvent accessibility prediction tools, a coiled coil prediction tool, and the tertiary structure prediction tool HHpred [52].

5 A Case Study

As noted in the Abstract, computational methods for predicting intrinsic disorder from amino acid sequences have been used in many analyses leading to new biological insights. Here I briefly introduce our recent study which found that interaction between mostly disordered proteins occurs frequently in a human PPI network [53].

The motivation behind this study was to investigate relation between the function of a PPI and the flexibility of the two interacting proteins. The classical model of PPI is the key-and-keyhole model, in which a folded protein is complementarily bound to a folded partner. Several groups have recently suggested other models, in which IDRs are folded when they are bound to their targets. To analyze PPIs from the viewpoint of the flexibility of the two interacting proteins, we simply categorized PPIs into three types—those between disordered regions (DR-DRs), structured regions (STR-STRs), and disordered and structured regions (DR-STRs)—and addressed the question "which of the three interaction types is the one most prevalent in existing PPI networks?"

Predictions for all proteins in a human PPI network were carried out using POODLE, and statistical testing was used to determine whether or not the numbers of DR-DRs, STR-STRs, and DR-STRs in the PPI network were significantly different from those in random networks. More precisely, we first extracted from the original PPI network a subnetwork comprising all proteins predicted to be either mostly disordered or mostly structured and counted in that subnetwork the interactions between (a) mostly disordered proteins, (b) mostly structured proteins, and (c) a mostly disordered protein and a mostly structured protein. We simply regarded as (a) DR-DRs, (b) STR-STRs, and (c) DR-STRs. Then we generated 1,000 random networks each of which had the same number of mostly disordered proteins, mostly structured proteins, and PPIs as the subnetwork but in which the interactions between two proteins were randomly given. DR-DRs, STR-STRs, and DR-STRs in each random network were counted in the same manner. We compared the numbers of DR-DRs, STR-STRs, and DR-STRs in the subnetwork against those in the 1,000 random networks and estimated *p*-values by using a two-tailed test of *z*-score. An overview of the method is shown in Fig. 5.

This analysis revealed that the occurrence of DR-DRs was significantly frequent and the occurrence of DR-STRs was significantly infrequent. We also found that this propensity was much stronger in interactions between non-hub proteins. Similar analyses were performed to determine whether or not a part of the human PPI network that is involved in a specific GO term is enriched in DR-DRs. This analysis yielded results demonstrating that DR-DRs frequently occur in cellular processes, regulation, and especially in metabolic processes.

The biological insight derived in the above analysis has been cited in a wide range of studies, including those related to diseases [54, 55], and drug discovery [56–58], as well as genome-wide functional analyses using IDRs predictions [59–61].

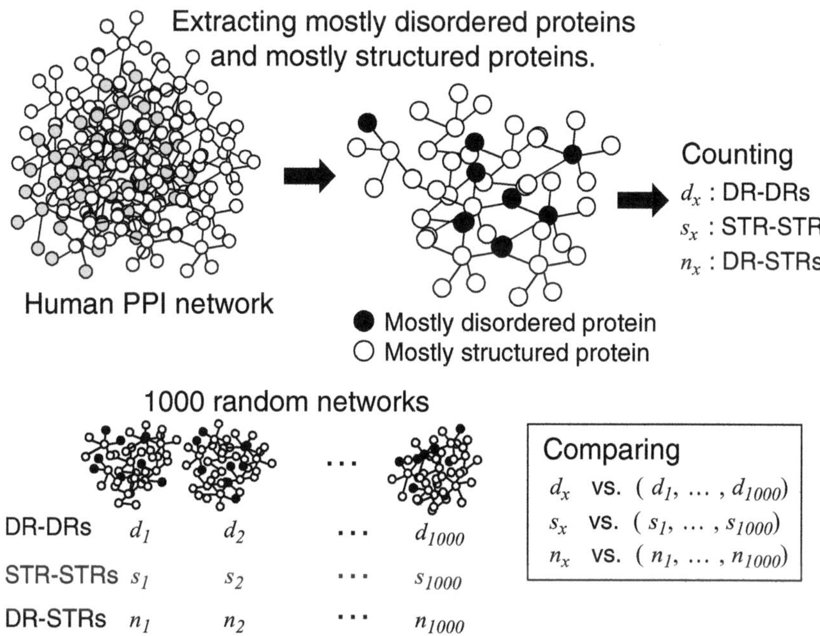

Fig. 5 Overview of a statistical analysis of IDRs on a human PPI network

Acknowledgments

I thank the co-developers of the POODLE series: Dr. Shuichi Hirose, Dr. Satoru Kanai, Dr. Yoichi Muraoka, and Dr. Tamotsu Noguchi. I also thank Dr. Kentaro Tomii, and Dr. Hiroyuki Toh for fruitful discussions.

References

1. Wright PE, Dyson HJ (1999) Intrinsically unstructured proteins: re-assessing the protein structure-function paradigm. J Mol Biol 293: 321–331
2. Dunker AK, Brown CJ, Lawson JD et al (2002) Intrinsic disorder and protein function. Biochemistry 41:6573–6582
3. Tompa P (2002) Intrinsically unstructured proteins. Trends Biochem Sci 27:527–533
4. Uversky VN (2002) Natively unfolded proteins: a point where biology waits for physics. Protein Sci 11:739–756
5. Tompa P (2005) The interplay between structure and function in intrinsically unstructured proteins. FEBS Lett 579:3346–3354
6. He B, Wang K, Liu Y et al (2009) Predicting intrinsic disorder in proteins: an overview. Cell Res 19:929–949
7. Deng X, Eickholt J, Cheng J (2012) A comprehensive overview of computational protein disorder prediction methods. Mol Biosyst 8:114–121
8. Longhi S, Receveur-Brechot V, Karlin D et al (2003) The C-terminal domain of the measles virus nucleoprotein is intrinsically disordered and folds upon binding to the C-terminal moiety of the phosphoprotein. J Biol Chem 278:18638–18648
9. Spinola-Amilibia M, Rivera J, Ortiz-Lombardia M et al (2011) The structure of BRMS1

nuclear export signal and SNX6 interacting region reveals a hexamer formed by antiparallel coiled coils. J Mol Biol 411:1114–1127

10. Reingewertz TH, Shalev DE, Sukenik S et al (2011) Mechanism of the interaction between the intrinsically disordered C-terminus of the pro-apoptotic ARTS protein and the Bir3 domain of XIAP. PLoS One 6:e24655
11. Mcdonald CB, Balke JE, Bhat V et al (2012) Multivalent binding and facilitated diffusion account for the formation of the Grb2-Sos1 signaling complex in a cooperative manner. Biochemistry 51:2122–2135
12. Mcdonald CB, Bhat V, Mikles DC et al (2012) Bivalent binding drives the formation of the Grb2-Gab1 signaling complex in a noncooperative manner. FEBS J 279:2156–2173
13. Khan H, Cino EA, Brickenden A et al (2013) Fuzzy complex formation between the intrinsically disordered prothymosin alpha and the Kelch domain of Keap1 involved in the oxidative stress response. J Mol Biol 425(6): 1011–1027
14. Ward JJ, Sodhi JS, Mcguffin LJ et al (2004) Prediction and functional analysis of native disorder in proteins from the three kingdoms of life. J Mol Biol 337:635–645
15. Motono C, Nakata J, Koike R et al (2011) SAHG, a comprehensive database of predicted structures of all human proteins. Nucleic Acids Res 39:D487–D493
16. Linding R, Russell RB, Neduva V et al (2003) GlobPlot: exploring protein sequences for globularity and disorder. Nucleic Acids Res 31:3701–3708
17. Dunker AK, Obradovic Z, Romero P et al (2000) Intrinsic protein disorder in complete genomes. Genome Inform Ser Workshop Genome Inform 11:161–171
18. Shimizu K, Muraoka Y, Hirose S et al (2007) Predicting mostly disordered proteins by using structure-unknown protein data. BMC Bioinforma 8:78
19. Dunker AK, Silman I, Uversky VN et al (2008) Function and structure of inherently disordered proteins. Curr Opin Struct Biol 18: 756–764
20. Minezaki Y, Homma K, Kinjo AR et al (2006) Human transcription factors contain a high fraction of intrinsically disordered regions essential for transcriptional regulation. J Mol Biol 359:1137–1149
21. Dunker AK, Cortese MS, Romero P et al (2005) Flexible nets. The roles of intrinsic disorder in protein interaction networks. FEBS J 272:5129–5148
22. Dosztanyi Z, Chen J, Dunker AK et al (2006) Disorder and sequence repeats in hub proteins and their implications for network evolution. J Proteome Res 5:2985–2995
23. Haynes C, Oldfield CJ, Ji F et al (2006) Intrinsic disorder is a common feature of hub proteins from four eukaryotic interactomes. PLoS Comput Biol 2:e100
24. Singh GP, Ganapathi M, Dash D (2007) Role of intrinsic disorder in transient interactions of hub proteins. Proteins 66:761–765
25. Uversky VN, Gillespie JR, Fink AL (2000) Why are "natively unfolded" proteins unstructured under physiologic conditions? Proteins 41:415–427
26. Linding R, Jensen LJ, Diella F et al (2003) Protein disorder prediction: implications for structural proteomics. Structure 11:1453–1459
27. Jones DT, Ward JJ (2003) Prediction of disordered regions in proteins from position specific score matrices. Proteins 53(Suppl 6):573–578
28. Prilusky J, Felder CE, Zeev-Ben-Mordehai T et al (2005) FoldIndex: a simple tool to predict whether a given protein sequence is intrinsically unfolded. Bioinformatics 21:3435–3438
29. Dosztanyi Z, Csizmok V, Tompa P et al (2005) The pairwise energy content estimated from amino acid composition discriminates between folded and intrinsically unstructured proteins. J Mol Biol 347:827–839
30. Shimizu K, Hirose S, Noguchi T (2007) POODLE-S: web application for predicting protein disorder by using physicochemical features and reduced amino acid set of a position-specific scoring matrix. Bioinformatics 23: 2337–2338
31. Hirose S, Shimizu K, Kanai S et al (2007) POODLE-L: a two-level SVM prediction system for reliably predicting long disordered regions. Bioinformatics 23:2046–2053
32. Shimizu K, Muraoka Y, Hirose S et al (2005) Feature selection based on physicochemical properties of redefined N-term region and C-term regions for predicting disorder. In: Proceedings of 2005 IEEE symposium on computational intelligence in bioinformatics and computational biology, pp 262–267
33. Peng K, Radivojac P, Vucetic S et al (2006) Length-dependent prediction of protein intrinsic disorder. BMC Bioinforma 7:208
34. Ward JJ, Mcguffin LJ, Bryson K et al (2004) The DISOPRED server for the prediction of protein disorder. Bioinformatics 20:2138–2139
35. Dosztanyi Z, Csizmok V, Tompa P et al (2005) IUPred: web server for the prediction of intrinsically unstructured regions of proteins based

on estimated energy content. Bioinformatics 21:3433–3434

36. Yang ZR, Thomson R, Mcneil P et al (2005) RONN: the bio-basis function neural network technique applied to the detection of natively disordered regions in proteins. Bioinformatics 21:3369–3376
37. Ishida T, Kinoshita K (2007) PrDOS: prediction of disordered protein regions from amino acid sequence. Nucleic Acids Res 35: W460–W464
38. Hirose S, Shimizu K, Noguchi T (2010) POODLE-I: disordered region prediction by integrating POODLE series and structural information predictors based on a workflow approach. In Silico Biol 10:185–191
39. Xue B, Dunbrack RL, Williams RW et al (2010) PONDR-FIT: a meta-predictor of intrinsically disordered amino acids. Biochim Biophys Acta 1804:996–1010
40. Ishida T, Kinoshita K (2008) Prediction of disordered regions in proteins based on the meta approach. Bioinformatics 24:1344–1348
41. Kozlowski LP, Bujnicki JM (2012) MetaDisorder: a meta-server for the prediction of intrinsic disorder in proteins. BMC Bioinforma 13:111
42. Bordoli L, Kiefer F, Schwede T (2007) Assessment of disorder predictions in CASP7. Proteins
43. Noivirt-Brik O, Prilusky J, Sussman JL (2009) Assessment of disorder predictions in CASP8. Proteins 77(Suppl 9):210–216
44. Monastyrskyy B, Fidelis K, Moult J et al (2011) Evaluation of disorder predictions in CASP9. Proteins 79(Suppl 10):107–118
45. Sakharkar MK, Sakharkar KR, Chow VT (2009) Human genomic diversity, viral genomics and proteomics, as exemplified by human papillomaviruses and H5N1 influenza viruses. Hum Genomics 3:320–331
46. Scotti C, Olivieri C, Boeri L et al (2011) Bioinformatic analysis of pathogenic missense mutations of activin receptor like kinase 1 ectodomain. PLoS One 6:e26431
47. Morita M, Saito S, Ikeda K et al (2009) Structural bases of GM1 gangliosidosis and Morquio B disease. J Hum Genet 54: 510–515
48. Mcguffin LJ, Bryson K, Jones DT (2000) The PSIPRED protein structure prediction server. Bioinformatics 16:404–405
49. Cuff JA, Barton GJ (2000) Application of multiple sequence alignment profiles to improve protein secondary structure prediction. Proteins 40:502–511
50. Adamczak R, Porollo A, Meller J (2005) Combining prediction of secondary structure and solvent accessibility in proteins. Proteins 59:467–475
51. Lupas A, Van Dyke M, Stock J (1991) Predicting coiled coils from protein sequences. Science 252:1162–1164
52. Soding J, Biegert A, Lupas AN (2005) The HHpred interactive server for protein homology detection and structure prediction. Nucleic Acids Res 33:W244–W248
53. Shimizu K, Toh H (2009) Interaction between intrinsically disordered proteins frequently occurs in a human protein-protein interaction network. J Mol Biol 392(5):1253–1265
54. Das S, Mukhopadhyay D (2011) Intrinsically unstructured proteins and neurodegenerative diseases: conformational promiscuity at its best. IUBMB Life 63:478–488
55. Manich G, Mercader C, Del Valle J et al (2011) Characterization of amyloid-beta granules in the hippocampus of SAMP8 mice. J Alzheimers Dis 25:535–546
56. Khan SH, Ahmad F, Ahmad N et al (2011) Protein-protein interactions: principles, techniques, and their potential role in new drug development. J Biomol Struct Dyn 28: 929–938
57. Wang J, Cao Z, Zhao L et al (2011) Novel strategies for drug discovery based on intrinsically disordered proteins (IDPs). Int J Mol Sci 12:3205–3219
58. Liu J, Li S, Dunker AK et al (2012) Molecular profiling: an essential technology enabling personalized medicine in breast cancer. Curr Drug Targets 13:541–554
59. Patil A, Kinoshita K, Nakamura H (2010) Hub promiscuity in protein-protein interaction networks. Int J Mol Sci 11:1930–1943
60. Nilsson J, Grahn M, Wright AP (2011) Proteome-wide evidence for enhanced positive Darwinian selection within intrinsically disordered regions in proteins. Genome Biol 12:R65
61. Nido GS, Mendez R, Pascual-Garcia A et al (2012) Protein disorder in the centrosome correlates with complexity in cell types number. Mol Biosyst 8:353–367

Chapter 11

Prediction of Intrinsic Disorder in Proteins Using MFDp2

Marcin J. Mizianty, Vladimir Uversky, and Lukasz Kurgan

Abstract

Intrinsically disordered proteins (IDPs) are either entirely disordered or contain disordered regions in their native state. IDPs were found to be abundant across all kingdoms of life, particularly in eukaryotes, and are implicated in numerous cellular processes. Experimental annotation of disorder lags behind the rapidly growing sizes of the protein databases and thus computational methods are used to close this gap and to investigate the disorder. MFDp2 is a novel webserver for accurate sequence-based prediction of protein disorder which also outputs well-described sequence-derived information that allows profiling the predicted disorder. We conveniently visualize sequence conservation, predicted secondary structure, relative solvent accessibility, and alignments to chains with annotated disorder. The webserver allows predictions for multiple proteins at the same time, includes help pages and tutorial, and the results can be downloaded as text-based (parsable) file. MFDp2 is freely available at http://biomine.ece.ualberta.ca/MFDp2/.

Key words Intrinsic disorder, Intrinsically disordered protein, Intrinsically disordered region, Prediction

1 Introduction

The intrinsically disordered proteins (IDPs), also called intrinsically unstructured or natively unfolded, are either entirely disordered or contain disordered regions in their native state. These highly flexible polypeptide chains form an ensemble of conformational states in vivo with no stable tertiary structure [1]. Regions of IDP can exist as unfolded chains or molten globules with well-developed secondary structure and they often function through transition between differently folded states [2].

Interest in IDPs continues to grow as these proteins were found to be implicated in numerous cellular processes including signal transduction, transcriptional regulation, and translation [3], cell death regulation [4], protein–DNA [5] and protein–protein [6] interactions. The disorder was demonstrated to play a role in several human diseases [7, 8], including AIDS [9], cancer [10], cardiovascular disease [11], neurodegenerative

Daisuke Kihara (ed.), *Protein Structure Prediction*, Methods in Molecular Biology, vol. 1137, DOI 10.1007/978-1-4939-0366-5_11,

diseases [12, 13], genetic diseases [14], and amyloidosis [15]. Moreover, IDPs have been shown to be abundant in across various organisms [16–20].

Prediction of disorder from protein sequences provides means to annotate and functionally characterize disorder for the ever growing number of protein chains. The MFDp2 [21] webserver can be used to predict and analyze per-residue intrinsic disorder probability given a protein sequence. Although many alternative disorder predictors are available [22–24], recent evaluation shows that MFDp2 is among the most accurate predictors [21]. Moreover, the MFDp2 webserver outputs a well-described profile that visualizes certain relevant structural and functional aspects of the predicted disorder. Our method utilizes per-residue predictions generated by MFDp [25], which are corrected to match disorder content predicted by DisCon [26]. Predictions are also filtered using post-processing filters and are enriched with alignment to known disorder regions available in PDB [27] and a curated repository of IDPs, Disprot [28], which improves predictions quality. MFDp2 is available as an easy to use webserver that not only predicts the disorder, but it also provides and conveniently visualizes per-residue conservation, list of aligned disordered regions from our template database, and several predicted structural characteristics of the input protein, such as secondary structure (predicted by PSIPRED [29]) and relative solvent accessibility (predicted by Real-SPINE3 [30]). This additional information is useful to profile the predicted disorder, e.g., to gain insights into how the disorder was predicted (from alignment, from MFDp, etc.) and to characterize the underlying structural properties (conservation, solvent accessibility, etc.). The webserver allows predictions to be downloaded as parsable text files, which facilitates downstream analysis. For convenience, these text files can be downloaded in two formats: as comma-separable CSV and/or FASTA. The webserver allows for analysis of sets of up to 100 proteins.

2 Materials

The webserver is designed to be simple to use. The submission page includes a text field where up to 100 protein sequences in FASTA format can be pasted and another text field for a user e-mail. Server also provides an option to submit proteins in FASTA-formatted file. The e-mail is optional and is used to send notification once the predictions are completed. The results are also shown and linked directly in a browser window after the prediction process starts. The help and tutorial page can be accessed at the top of the main webserver page. It explains how to use the webserver and provides detailed explanations on how to read the results. Individual subsections of the help and tutorial page are hyperlinked within this page and from the pages that the user encounters when interacting with the server to ease finding of this information. The explanations are supplemented with annotated screenshots. The "?" buttons are placed thorough all webserver pages next to the sections which may require explanation. These buttons implement direct hyperlinks to the help and hints related to the corresponding section/task.

The MFDp2 uses other programs to perform and to visualize predictions. Our method predicts the disorder utilizing predictions generated by MFDp and DisCon, as well as alignment using PSI-BLAST [31]. The profile that accompanies the prediction includes information about residues conservation, protein secondary structure predicted by PSIPRED, and solvent accessibility predicted by Real-SPINE3.

The webserver, which includes help pages and tutorial, is freely available at http://biomine.ece.ualberta.ca/MFDp2/.

3 Methods

3.1 Running MFDp2

Three easy steps should be followed to use the MFDp2 webserver (step numbers are given in Fig. 1):

1. Copy and paste protein sequences list in the FASTA format into text field or upload FASTA-formatted file (an "Example" button may be used to see an example input of the FASTA format) (*see* **Notes 1** and **2**).
2. Provide e-mail address (optional). If e-mail is provided, a notification e-mail will be sent once the results are ready. The notification will include a web address where the results are stored (*see* **Note 3**).
3. Click "Run MFDp2" button to start the predictions (*see* **Note 4**).

Once the prediction is finished, the user is directed to the results that are available through two web pages: "results summary page" and "detailed results page".

3.2 Results Summary Page

This page provides overview of predictions made by MFDp2 webserver for all submitted proteins and contains links to more detailed per-protein pages (*see* Subheading 3.3). Following options and information are available (numbered options are shown in Fig. 2):

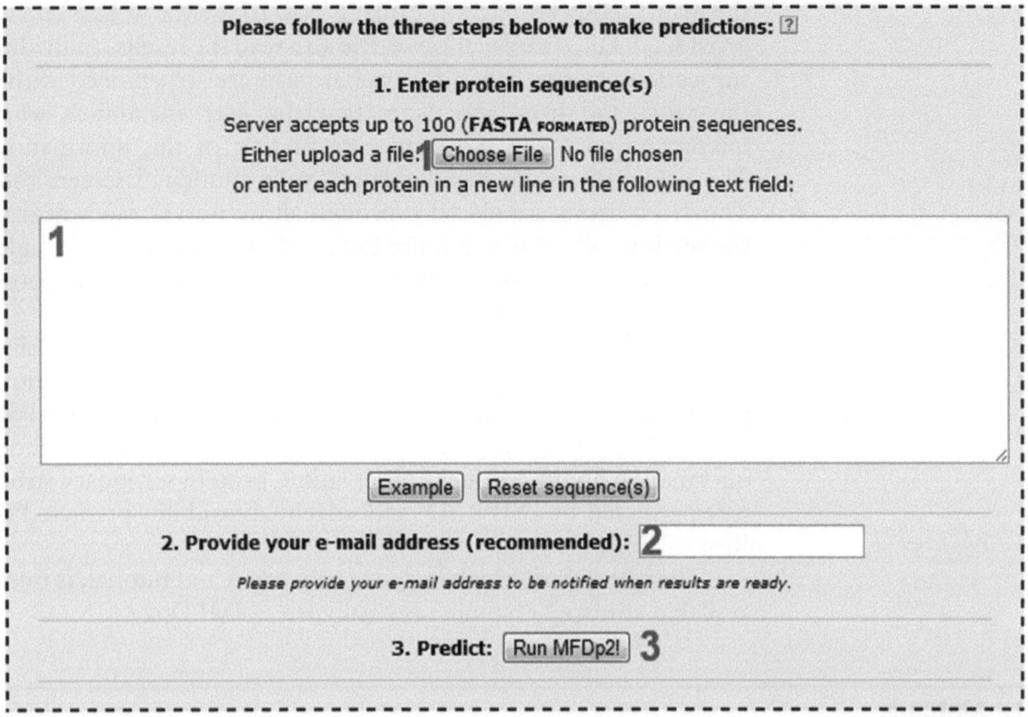

Please follow the three steps below to make predictions:

1. Enter protein sequence(s)

Server accepts up to 100 (FASTA FORMATED) protein sequences.
Either upload a file: 1 Choose File No file chosen
or enter each protein in a new line in the following text field:

1

Example Reset sequence(s)

2. Provide your e-mail address (recommended): 2

Please provide your e-mail address to be notified when results are ready.

3. Predict: Run MFDp2! 3

Fig. 1 Screenshot of MFDp2 input form on the main webserver page. The *large red numbers* annotate major elements on this page

1. Predictions may be downloaded as .csv or .fasta file (*see* Subheading 3.4).
2. Summary of results shows brief statistics of the predicted disorder followed by per-residue binary disorder prediction for each submitted protein (*see* **Note 5**).
3. More detailed predictions for a given protein can be accessed by clicking on the protein name or sequence.

3.3 Detailed Results Page

This page provides more detailed information about the predicted disorder for a given protein. The following options and information are available (numbered options are annotated in Fig. 3):

1. The menu on the top of the page contains links that the user may utilize to navigate this page.
2. Overview includes brief statistics concerning the predicted disorder, such as disorder content, number of disordered regions, and number of templates with aligned disorder regions, followed by the per-residue binary disorder prediction (*see* **Note 5**).
3. Per-residue disorder profiles. The profile includes conveniently visualized information concerning per-residue conservation (denoted "R.Ent"), predicted secondary structure (denoted

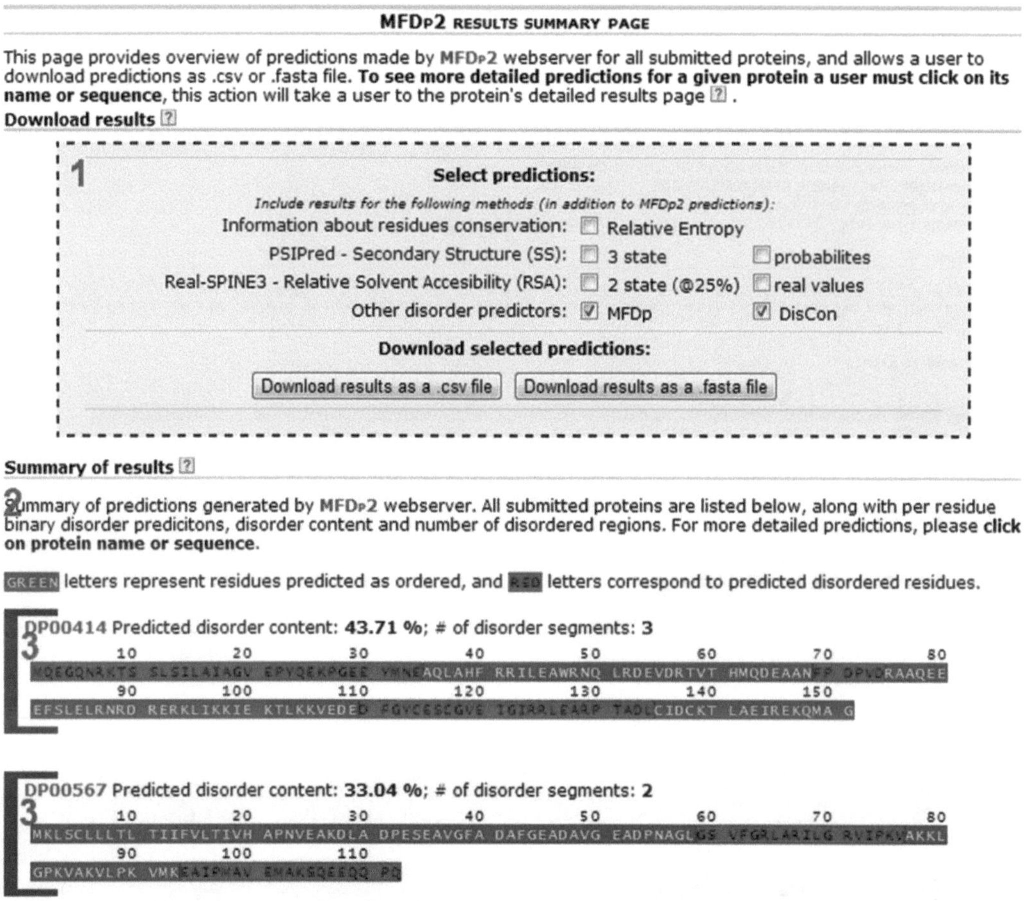

MFDp2 RESULTS SUMMARY PAGE

This page provides overview of predictions made by MFDp2 webserver for all submitted proteins, and allows a user to download predictions as .csv or .fasta file. **To see more detailed predictions for a given protein a user must click on its name or sequence**, this action will take a user to the protein's detailed results page.

Download results

1

Select predictions:

Include results for the following methods (in addition to MFDp2 predictions):

Information about residues conservation: Relative Entropy

PSIPred - Secondary Structure (SS): 3 state; probabilites

Real-SPINE3 - Relative Solvent Accesibility (RSA): 2 state (@25%); real values

Other disorder predictors: MFDp; DisCon

Download selected predictions:

Download results as a .csv file | Download results as a .fasta file

Summary of results

2

Summary of predictions generated by MFDp2 webserver. All submitted proteins are listed below, along with per residue binary disorder predicitons, disorder content and number of disordered regions. For more detailed predictions, please **click on protein name or sequence**.

GREEN letters represent residues predicted as ordered, and RED letters correspond to predicted disordered residues.

3

DP00414 Predicted disorder content: **43.71 %**; # of disorder segments: **3**

MQEGQNRKTS SLSILAIAGV EPYQEKPGEE YMNEAQLAHF RRILEAWRNQ LRDEVDRTVT HMQDEAANFP DPVDRAAQEE
EFSLELRNRD RERKLIKKIE KTLKKVEDED FGYCESCGVE IGIRRLEARP TADLCIDCKT LAEIREKQMA G

3

DP00567 Predicted disorder content: **33.04 %**; # of disorder segments: **2**

MKLSCLLLTL TIIFVLTIVH APNVEAKDLA DPESEAVGFA DAFGEADAVG EADPNAGLGS VFGRLARILG RVIPKVAKKL
GPKVAKVLPK VMKEAIPMAV EMAKSQEEQQ PQ

Fig. 2 Screenshot of MFDp2 results summary page. The *large red numbers* annotate major elements on this page

"SS"), disorder profiles predicted with MFDp (denoted "MFDp"), predicted relative solvent accessibility (denoted "RSA"), and aligned disorder segments (denoted "BLAST") (*see* **Notes 6** and 7). The profile is color coded to ease the interpretation, where a spectrum of colors between red and green (except for the conservation) corresponds to the bias towards disordered and ordered conformations, respectively. Conservation information is color coded from white, corresponding to the least conserved residues, to black for the most conserved amino acids.

4. Segments section shows a set of basic statistics including length and position of the disordered segment in the sequence.
5. Alignments section lists all template proteins which were used to generate prediction (*see* **Note 8**). Beside the basic alignment statistics, the alignment itself is presented together with the annotated predicted disorder and actual disorder label for the query and subject proteins, respectively.

MFDp2 DETAILED RESULTS PAGE

Detailed results for MFDp2 webserver. Go back to OVERVIEW OF PREDCITIONS FOR ALL SUBMITED PROTEINS.

DP00414

The results below are divided into three sections:

1
- **OVERVIEW** - Overall information about predicted disorder
- **PROBABILITY** - Per residue probability graph
- **DISORDER SEGMENTS** - Detailed information about each predicted disorder segment
- **DISORDER ALIGNMENTS** - Detailed information about each aligned disorder

Overview

DP00414 is 151 residues long, with 66 residues (43.71%) predicted as disordered. The protein has 2 short (< 30 residues) disorder segments and 1 long (>= 30 residues) disorder segment. Moreover, protein was aligned to disorder regions from **1** TEMPLATE.

2

Amino acid sequence :

```
        10         20         30         40         50         60         70         80
MQEGQNRKTS SLSILAIAGV EPYQEKPGEE YMNEAQLAHF RRILEAWRNQ LRDEVDRTVT HMQDEAANFP DPVDRAAQEE
        90        100        110        120        130        140        150
EFSLELRNRD RERKLIKKIE KTLKKVEDED FGYCESCGVE IGIRRLEARP TADLCIDCKT LAEIREKQMA G
```

GREEN letters represent residues predicted as ordered, and RED letters correspond to predicted disordered residues.

Per residue disorder profiles

3

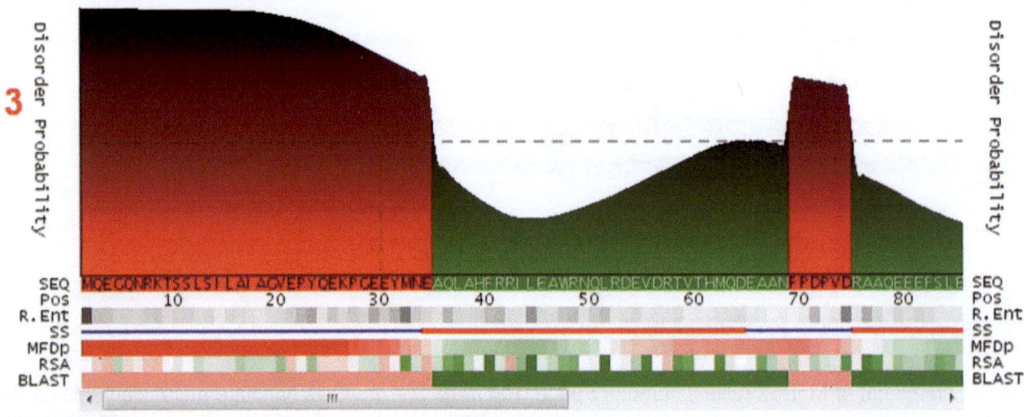

Segments

4

1. Segment 1 - Long (>= 30 residues) disordered segment
   Segment is located between positions 1 and 34 in the sequence.
   The segment is 34 residues long (22.52 % of the total sequence length).
2. Segment 2 - Short (< 30 residues) disordered segment
   Segment is located between positions 69 and 74 in the sequence.
   The segment is 6 residues long (3.97 % of the total sequence length).
3. Segment 3 - Short (< 30 residues) disordered segment
   Segment is located between positions 110 and 134 in the sequence.
   The segment is 25 residues long (16.56 % of the total sequence length).

Alignments

The query protein has some of its predicted disorder residues aligned to following proteins:

5

DP00414

```
     Link: DP00414
   Expect = 3.0E-86, Identities = 151/151 , Positives = 151/151
   Subject length: 151, position of subject alignment: 1-151
     Query length: 151, position of query alignment: 1-151
                   10        20        30        40        50        60        70        80
    Query MQEGQNRKTSSLSILAIAGVEPYQEKPGEEYMNEAQLAHFRRILEAWRNQLRDEVDRTVTHMQDEAANFPDPVDRAAQEEEFSLE
Alignment MQEGQNRKTSSLSILAIAGVEPYQEKPGEEYMNEAQLAHFRRILEAWRNQLRDEVDRTVTHMQDEAANFPDPVDRAAQEEEFSLE
  Subject MQEGQNRKTSSLSILAIAGVEPYQEKPGEEYMNEAQLAHFRRILEAWRNQLRDEVDRTVTHMQDEAANFPDPVDRAAQEEEFSLE
```

Fig. 3 Screenshot of MFDp2 detailed results page. The *large red numbers* annotate major elements on this page

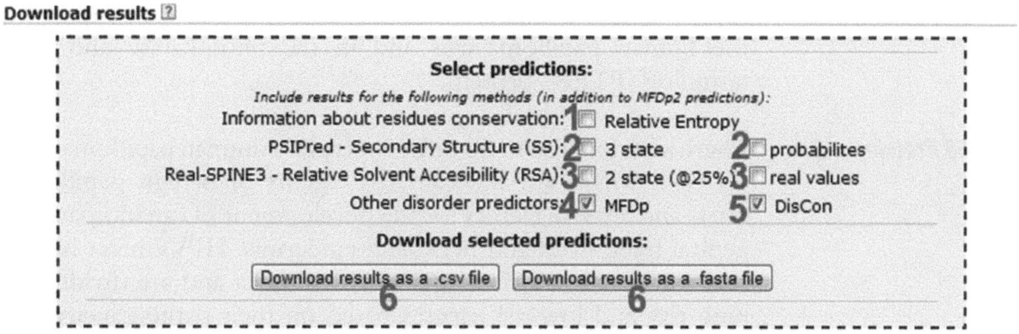

Download results

Select predictions:

Include results for the following methods (in addition to MFDp2 predictions):

Information about residues conservation: 1 Relative Entropy

PSIPred - Secondary Structure (SS): 2 3 state 2 probabilites

Real-SPINE3 - Relative Solvent Accesibility (RSA): 3 2 state (@25%) 3 real values

Other disorder predictors: 4 MFDp 5 DisCon

Download selected predictions:

Download results as a .csv file 6 Download results as a .fasta file 6

Fig. 4 Screenshot of the form that offers options concerning downloading of the predictions

3.4 Downloading the Predictions

This form, available on the results summary page, allows a user to download the predictions. The resulting file always contains the protein sequence and the MFDp2 predictions, including both per-residue probabilities and binary predictions. Following options, which are numbered in Fig. 4, are available:

1. This box should be selected to include information about the per-residue conservation expressed by relative entropy [32]. The entropy values are calculated using weighted observed percentages (WOP) matrix generated by PSI-BLAST.
2. These boxes should be selected to include Secondary Structure (SS) predicted by PSIPRED (both per-residue probabilities and three state predictions are available).
3. These boxes should be selected to include RSA predicted by Real-SPINE3 (both per-residue probabilities and binary predictions are available).
4. This box should be selected to include disorder predicted by MFDp (predecessor of MFDp2) (both per-residue probabilities and binary predictions will be added).
5. This box should be selected to include disorder content predicted by DisCon.
6. The selected set of predictions can be downloaded in either .csv (*see* **Note 9**) or .fasta (*see* **Note 10**) format.

4 Case Studies

MFDp, which is MFDp2's predecessor, has been used in a number of studies that characterize abundance and functional roles of intrinsic disorder in HIV-1 proteome [9], histone proteins [5], and proteins involved in the programmed cell death [4]. MFDp2 was not yet utilized in a similar fashion since it was published only recently. To this end, we present results of two case studies that apply MFDp2 to analyze intrinsic disorder in the E6 protein from the human papillomavirus and in the phosphatase and tensin homolog (PTEN) protein.

4.1 E6 Protein

There are more than 100 different types of human papillomaviruses (HPVs), which are the causative agents of benign papillomas/warts and are risk factors for the development of carcinomas of the genital tract, head and neck, and epidermis. HPVs infect mucosal and cutaneous stratified squamous epithelia and are divided into high-risk and low-risk viruses based on their pathogenicity [33]. For example, HPV-6 and HPV-11 DNAs are the predominant types found in genital warts (condyloma accuminata), whereas HPV-16 and HPV-18 DNAs are the predominantly associated with cervical carcinoma. Thus, HPV-6 and HPV-11 are referred to as low-risk (with respect to the cervical cancer) and HPV-16 and HPV-18 are referred to as the high-risk types.

E6 is one of the two oncoproteins of HPV that are responsible for HPV-mediated malignant cell progression, leading ultimately to an invasive carcinoma. Protein E6 acts as an oncoprotein in the high-risk HPVs and promotes tumorigenesis by stimulating cellular degradation of the tumor suppressor p53 via formation of a trimeric complex comprising E6, p53, and the cellular ubiquitination enzyme E6AP [34, 35]. In addition, E6 displays numerous activities unrelated to p53. These include but are not limited to the recognition of a variety of other cellular proteins, such as transcription coactivators p300/CBP [36, 37] and ADA3 [38]; transcription factors c-Myc [39] and IRF3 [40]; replication protein hMCM7 [41]; DNA repair proteins MGMT [42]; protein kinases PKN [43] and Tyk2 [44]; Rap-GTPase activating protein E6TP1 [45]; tumor necrosis factor receptor TNF-R1 [46]; apoptotic protein Bak [47]; clathrin-adaptor complex AP-1 [48]; focal adhesion component paxillin [48] calcium-binding proteins E6BP [49], and fibulin-1 [50]; and several members of the PDZ protein family including hDLG [51], hScrib [52], MAGI-1 [53], and MUPP1 [54]. Furthermore, E6 activates or represses several cellular or viral transcription promoters [40, 55–57], e.g., it induces transcriptional activation of the gene encoding the retrotranscriptase of human telomerase [58, 59]. In addition, E6 recognizes four-way DNA junctions [60, 61]. The function of the low-risk HPV E6 is less well studied. However, the low-risk E6 lacks a number of activities which correlate with the oncogenic activity of the high-risk HPV E6. For example, the low-risk E6 does not bind PDZ proteins [51] or E6TP1 [45] and does not target p53 for degradation [34, 62]. Like the high-risk E6, the low-risk E6 binds MCM7 [41] and Bak [47] and inhibits p300 acetylation of p53 [63].

Sequence alignments of E6 proteins from numerous HPV subtypes suggested the presence of two zinc-binding motifs, which are 37 residues long regions that contain four cysteines distributed in

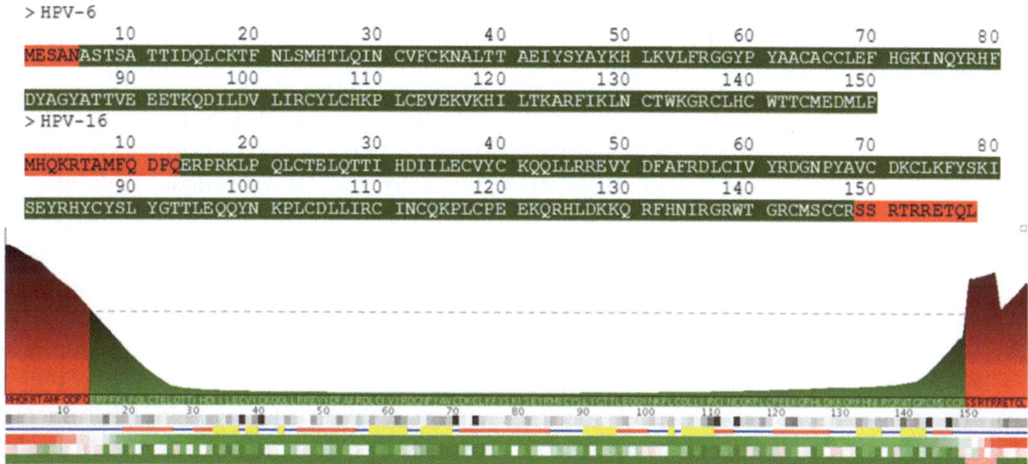

Fig. 5 MFDp2 predictions for the HPV-6 (UniProt ID: P06462) and HPV-16 (UniProt ID: P03126) proteins. Ordered and disordered residues are shown on *green* and *red* background, respectively. The graphical representation of predicted disorder along the HPV-16 protein sequence is shown below (*see* **Note 10**)

a CxxC-(29x)-CxxC motif [64]. The sequence of E6 protein can be divided into five regions [65–68]: the N-terminal tail (residues 1–36), the N-terminal zinc-binding motif (residues 37–73), the linker region (residues 74–110), the C-terminal zinc-binding motif (residues 110–146), and the C-terminal tail (residues 147–158) (using the 158-residue numbering of HPV-16 protein E6). Based on the now available structural data on the N- and C-terminal domains of E6 is has been concluded that this protein contains two well-structured regions that correspond to functional domains (residues 12–71 and 80–143) and three unstructured fragments, N-tail (residues 1–11), C-tail (residues 144–153), and the interdomain linker (residues 72–80) [69].

Earlier, based on the bioinformatics analysis of proteins from the low- and high-risk HPVs it has been concluded that high-risk HPVs are characterized by the increased amount of intrinsic disorder in transforming proteins E6 and E7 [70, 71]. In agreement with these earlier studies, our analysis using MFDp2 revealed the noticeable difference in the disorder levels of E6 proteins from HPV-6 (3.3 %) and HPV-16 (14.6 %). The most disordered parts of the E6 from HPV-16 are its N- and C-terminal tails (*see* Fig. 5). Since the major structural difference between the E6 proteins from the low- and high-risk HPVs is the presence of disordered tails in the high-risk HPV proteins, and since the high-risk E6 proteins are characterized by a broader functional spectrum, it is tempting to hypothesize that the higher binding promiscuity of the E6 proteins from high-risk HPVS is due to the intrinsically disordered nature of their N- and C-terminal regions.

4.2 PTEN Protein

PTEN, a 403 amino acid protein/lipid phosphatase, is the second most frequently mutated tumor suppressor after p53 [72]. PTEN acts as a dual-specificity protein phosphatase, dephosphorylating tyrosine-, serine-, and threonine-phosphorylated proteins and also functions as a lipid phosphatase that converts phosphatidylinositol (3–5)-triphosphate (PtdIns(3–5)P3 or PIP3) to phosphatidylinositol 4,5-bisphosphate (PtdIns(4,5)P2 or PIP2). PTEN regulates the Phosphoinositide 3-Kinase/Akt/mammalian Target Of Rapamycin (PI3K/Akt/mTOR) pathway involved in oncogenic signaling, cell proliferation, survival and apoptosis, which are under the control of several growth factors [73]. Its protein phosphatse activity is under investigation and it was recently shown that PTEN autodephosphorylates itself utilizing its protein phosphatase activity [74]. Within the nucleus, PTEN maintains chromosomal stability during cell division [75]. PTEN loss causes uncontrolled cell proliferation and accumulation of mutations in cells, causing cancer. Indeed, deficiency and dysregulation of PTEN drives endometrial, prostate and brain cancers, and causes neurological defects [76–79].

The lipid phosphatase activity of PTEN is modulated via membrane association [80]. The active form of PTEN anchors to the plasma membrane via its PIP2 binding module (PBM) and C2 domain, providing conformational accessibility to the catalytic phosphatase domain that converts PIP3 to PIP2 [80]. Cancer causing mutations in PTEN may occur within or outside of the catalytic domain; mutations of the latter type inhibit PTEN function by preventing its membrane association [81].

Crystal structure of the central fragment of PTEN (amino acid position 7–353) was determined [82]. In spite of many attempts, the structure of three regions, i.e., the N-terminus (residues 1–13), the CBR3 loop (residues 280–314), and the C-terminal tail (residues 354–403), remains undetermined due to their highly dynamic nature [82]. Of particular interest is the C-tail which has been recently found to regulate PTEN intra-molecular interactions that dictate its membrane association, function, and stability through multiple phosphorylation events mediated by several kinases [83–85]. In agreement with this structural data, computational analysis with MFDp2 revealed that this protein possesses 36 % disordered residues, *see* Fig. 6. In fact, PTEN is a hybrid protein that by prediction contains seven short (<30 residues) disordered segments and one long (>60 residues) C-terminally located disordered region. Importantly, most of the predicted disordered regions of PTEN (residues 1–17, 38–49, 73–78, 89–99, 112–118, 158–163, 167–192, and 341–403) correspond either to the terminal segments (residues 1–17 and 351–403) that were experimentally shown to be disordered or to the flexible loops (residues 40–48, 72–84, 91–98, and 160–169). As far as the CBR3 loop (residues 280–314) is concerned, this segment is predicted to have increased flexibility, since its disorder score is close to the 0.5 threshold.

Fig. 6 MFDp2 predictions for the PTEN protein (UniProt ID: P60484). Ordered and disordered residues are shown on *green* and *red background*, respectively. The graphical representation of predicted disorder along the PTEN protein sequence is shown below (*see* **Note 10**)

5 Notes

1. Server accepts up to 100 protein sequences.
2. Due to a limitation of one of the methods that is used to generate MFDp predictions (HHSearch), the sever sometimes cannot process neither extremely short (<15 residues) nor very long (>1,000 residues) protein chains. In the rare event when the server is unable to generate predictions, the results for proteins for which predictions are ready will be displayed, and proteins which were not predicted will have appropriate annotation informing about the unavailability of the results.
3. Direct hyperlink to the results is provided once the "Run MFDp2!" button is pressed. User should store this link for future reference. The same link is sent via e-mail, if the e-mail address was provided.
4. The MFDp2's execution time is approximately 5–15 min for an average size protein chain. The time is mostly determined by the runtime to run PSI-BLAST.
5. Green letters represent residues predicted as ordered and red letters correspond to the predicted disordered residues. The border around a given protein is also color coded based on its disorder content, green border indicates proteins with low disorder content, whereas red border indicates protein with high disorder content.
6. The predicted per-residue intrinsic disorder probabilities are also available in the raw form in the files that can be downloaded from the "Results summary" page.
7. The profile is color coded to ease the interpretation. Values of the abovementioned characteristics (conservation, secondary

structure, etc.) which are associated with disordered residues are shown in red, while values associated with order are shown in green. The profile includes the following information:

(a) SEQ—AA sequence, GREEN letters represent residues predicted as ordered, and RED letters correspond to the predicted disordered residues.

(b) Pos—enumerates residues positions in the sequence. Number is displayed every ten residues, the last digit of the number overlaps with the enumerated residue.

(c) R. Ent—per-residue conservation score expressed as relative entropy which is calculated using WOP matrix generated by PSI-BLAST

(d) SS—three state Secondary Structure (SS) predicted by PSIpred (colors correspond to: blue—coil, red—alpha helix, yellow—beta sheet).

(e) MFDp—disorder probability predicted by MFDp.

(f) RSA—values of the relative solvent accessibility predicted by Real-SPINE3.

(g) BLAST—probability of disorder assessed by PSI-BLAST alignment to the database with proteins with annotated disorder segments.

8. Aligned template's protein name is a clickable link that points to the PDB or DisProt entry for this protein.

9. In the .csv file each line starts with a user-given protein name, the type of information that the line provides and the corresponding information. These three fields are comma separated. Example .csv file follows:

DP00582,AA Sequence,Q,D,K,C,K,K,V,Y,E,...

DP00582,MFDp2 probabiliies,0.499,0.499,0.466,0.435,0.408, 0.386,0.366,0.348,0.33 1,...

DP00582,MFDp2 binary,0,0,0,0,0,0,0,0,0,...

10. Each .fasta file starts with a header that identifies format of the subsequent data, and then the data is outputted for each protein. Example .fasta file follows:

```
#File format:
# >Protein name
#AA Sequence
#MFDp2 probabilities—separated by comma
#MFDp2 binary
>DP00582
QDKCKKVYE...
0.499,0.499,0.466,0.435,0.408,0.386,0.366,0.348,0.331,...
000000000...
```

Acknowledgments

This work was supported by the Dissertation fellowship awarded by the University of Alberta to M.J.M. and by the Discovery grant from the Natural Sciences and Engineering Research Council of Canada to L.K.

References

1. Uversky VN, Gillespie JR, Fink AL (2000) Why are "natively unfolded" proteins unstructured under physiologic conditions? Proteins 41:415–427
2. Uversky VN (2002) Natively unfolded proteins: a point where biology waits for physics. Protein Sci 11:739–756
3. Dunker AK, Oldfield CJ, Meng J, Romero P, Yang JY, Chen JW, Vacic V, Obradovic Z, Uversky VN (2008) The unfoldomics decade: an update on intrinsically disordered proteins. BMC Genomics 9 Suppl 2:S1
4. Peng Z, Xue B, Kurgan L, Uversky VN (2013) Resilience of death: intrinsic disorder in proteins involved in the programmed cell death. Cell Death Differ 20(9):1257–1267
5. Peng Z, Mizianty MJ, Xue B, Kurgan L, Uversky VN (2012) More than just tails: intrinsic disorder in histone proteins. Mol Biosyst 8:1886–1901
6. Russell RB, Gibson TJ (2008) A careful disorderliness in the proteome: sites for interaction and targets for future therapies. FEBS Lett 582:1271–1275
7. Uversky VN, Oldfield CJ, Midic U, Xie H, Xue B, Vucetic S, Iakoucheva LM, Obradovic Z, Dunker AK (2009) Unfoldomics of human diseases: linking protein intrinsic disorder with diseases. BMC Genomics 10 Suppl 1:S7
8. Uversky VN, Oldfield CJ, Dunker AK (2008) Intrinsically disordered proteins in human diseases: introducing the D2 concept. Annu Rev Biophys 37:215–246
9. Xue B, Mizianty MJ, Kurgan LA, Uversky VN (2012) Protein intrinsic disorder as a flexible armor and a weapon of HIV-1. Cell Mol Life Sci 69(8):1211–1259
10. Iakoucheva LM, Brown CJ, Lawson JD, Obradovic Z, Dunker AK (2002) Intrinsic disorder in cell-signaling and cancer-associated proteins. J Mol Biol 323:573–584
11. Cheng Y, LeGall T, Oldfield CJ, Dunker AK, Uversky VN (2006) Abundance of intrinsic disorder in protein associated with cardiovascular disease. Biochemistry 45:10448–10460
12. Raychaudhuri S, Dey S, Bhattacharyya NP, Mukhopadhyay D (2009) The role of intrinsically unstructured proteins in neurodegenerative diseases. PLoS One 4:e5566
13. Uversky VN (2009) Intrinsic disorder in proteins associated with neurodegenerative diseases. Front Biosci 14:5188–5238
14. Midic U, Oldfield CJ, Dunker AK, Obradovic Z, Uversky VN (2009) Protein disorder in the human diseasome: unfoldomics of human genetic diseases. BMC Genomics 10 Suppl 1:S12
15. Uversky VN (2008) Amyloidogenesis of natively unfolded proteins. Curr Alzheimer Res 5:260–287
16. Ward JJ, Sodhi JS, McGuffin LJ, Buxton BF, Jones DT (2004) Prediction and functional analysis of native disorder in proteins from the three kingdoms of life. J Mol Biol 337:635–645
17. Xue B, Dunker AK, Uversky VN (2012) Orderly order in protein intrinsic disorder distribution: disorder in 3500 proteomes from viruses and the three domains of life. J Biomol Struct Dyn 30:137–149
18. Pancsa R, Tompa P (2012) Structural disorder in eukaryotes. PLoS One 7:e34687
19. Yan J, Mizianty MJ, Filipow PL, Uversky VN, Kurgan L (2013) RAPID: fast and accurate sequence-based prediction of intrinsic disorder content on proteomic scale. Biochim Biophys Acta 1834:1671–1680
20. Peng Z, Mizianty MJ, Kurgan L (2013) Genome-scale prediction of proteins with long intrinsically disordered regions. Proteins Struct Funct Bioinformatics
21. Mizianty MJ, Peng Z, Kurgan L (2013) MFDp2: accurate predictor of disorder in proteins by fusion of disorder probabilities, content and profiles. Intrinsically Disordered Proteins 1:13–22
22. Deng X, Eickholt J, Cheng J (2012) A comprehensive overview of computational protein disorder prediction methods. Mol Biosyst 8:114–121

23. Peng Z-L, Kurgan LA (2012) Comprehensive comparative assessment of in-silico predictors of disordered regions. Curr Protein Pept Sci 13:6–18
24. He B, Wang K, Liu Y, Xue B, Uversky VN, Dunker AK (2009) Predicting intrinsic disorder in proteins: an overview. Cell Res 19:929–949
25. Mizianty MJ, Stach W, Chen K, Kedarisetti KD, Disfani FM, Kurgan LA (2010) Improved sequence-based prediction of disordered regions with multilayer fusion of multiple information sources. Bioinformatics 26:i489–i496
26. Mizianty MJ, Zhang T, Xue B, Zhou Y, Dunker AK, Uversky VN, Kurgan LA (2011) In-silico prediction of disorder content using hybrid sequence representation. BMC Bioinformatics 12:245
27. Berman HM, Westbrook J, Feng Z, Gilliland G, Bhat TN, Weissig H, Shindyalov IN, Bourne PE (2000) The Protein Data Bank. Nucleic Acids Res 28:235–242
28. Sickmeier M, Hamilton JA, LeGall T, Vacic V, Cortese MS, Tantos A, Szabo B, Tompa P, Chen J, Uversky VN et al (2007) DisProt: the database of disordered proteins. Nucleic Acids Res 35:D786–D793
29. McGuffin LJ, Bryson K, Jones DT (2000) The PSIPRED protein structure prediction server. Bioinformatics 16:404–405
30. Faraggi E, Xue B, Zhou Y (2009) Improving the prediction accuracy of residue solvent accessibility and real-value backbone torsion angles of proteins by guided-learning through a two-layer neural network. Proteins 74:847–856
31. Altschul SF, Madden TL, Schäffer AA, Zhang J, Zhang Z, Miller W, Lipman DJ (1997) Gapped BLAST and PSI-BLAST: a new generation of protein database search programs. Nucleic Acids Res 25:3389–3402
32. Wang K, Samudrala R (2006) Incorporating background frequency improves entropy-based residue conservation measures. BMC Bioinformatics 7:385
33. Zur Hausen H (2002) Papillomaviruses and cancer: from basic studies to clinical application. Nat Rev Cancer 2:342–350
34. Scheffner M, Werness BA, Huibregtse JM, Levine AJ, Howley PM (1990) The E6 oncoprotein encoded by human papillomavirus types 16 and 18 promotes the degradation of p53. Cell 63:1129–1136
35. Scheffner M, Huibregtse JM, Vierstra RD, Howley PM (1993) The HPV-16 E6 and E6-AP complex functions as a ubiquitin-protein ligase in the ubiquitination of p53. Cell 75:495–505
36. Patel D, Huang SM, Baglia LA, McCance DJ (1999) The E6 protein of human papillomavirus type 16 binds to and inhibits co-activation by CBP and p300. EMBO J 18:5061–5072
37. Zimmermann H, Degenkolbe R, Bernard HU, O'Connor MJ (1999) The human papillomavirus type 16 E6 oncoprotein can down-regulate p53 activity by targeting the transcriptional coactivator CBP/p300. J Virol 73:6209–6219
38. Kumar A, Zhao Y, Meng G, Zeng M, Srinivasan S, Delmolino LM, Gao Q, Dimri G, Weber GF, Wazer DE et al (2002) Human papillomavirus oncoprotein E6 inactivates the transcriptional coactivator human ADA3. Mol Cell Biol 22:5801–5812
39. Gross-Mesilaty S, Reinstein E, Bercovich B, Tobias KE, Schwartz AL, Kahana C, Ciechanover A (1998) Basal and human papillomavirus E6 oncoprotein-induced degradation of Myc proteins by the ubiquitin pathway. Proc Natl Acad Sci USA 95:8058–8063
40. Ronco LV, Karpova AY, Vidal M, Howley PM (1998) Human papillomavirus 16 E6 oncoprotein binds to interferon regulatory factor-3 and inhibits its transcriptional activity. Genes Dev 12:2061–2072
41. Kukimoto I, Aihara S, Yoshiike K, Kanda T (1998) Human papillomavirus oncoprotein E6 binds to the C-terminal region of human minichromosome maintenance 7 protein. Biochem Biophys Res Commun 249:258–262
42. Srivenugopal KS, Ali-Osman F (2002) The DNA repair protein, O(6)-methylguanine-DNA methyltransferase is a proteolytic target for the E6 human papillomavirus oncoprotein. Oncogene 21:5940–5945
43. Gao Q, Kumar A, Srinivasan S, Singh L, Mukai H, Ono Y, Wazer DE, Band V (2000) PKN binds and phosphorylates human papillomavirus E6 oncoprotein. J Biol Chem 275:14824–14830
44. Li S, Labrecque S, Gauzzi MC, Cuddihy AR, Wong AH, Pellegrini S, Matlashewski GJ, Koromilas AE (1999) The human papilloma virus (HPV)-18 E6 oncoprotein physically associates with Tyk2 and impairs Jak-STAT activation by interferon-alpha. Oncogene 18:5727–5737
45. Gao Q, Srinivasan S, Boyer SN, Wazer DE, Band V (1999) The E6 oncoproteins of high-risk papillomaviruses bind to a novel putative GAP protein, E6TP1, and target it for degradation. Mol Cell Biol 19:733–744

46. Filippova M, Song H, Connolly JL, Dermody TS, Duerksen-Hughes PJ (2002) The human papillomavirus 16 E6 protein binds to tumor necrosis factor (TNF) R1 and protects cells from TNF-induced apoptosis. J Biol Chem 277:21730–21739
47. Thomas M, Banks L (1999) Human papillomavirus (HPV) E6 interactions with Bak are conserved amongst E6 proteins from high and low risk HPV types. J Gen Virol 80(Pt 6):1513–1517
48. Tong X, Boll W, Kirchhausen T, Howley PM (1998) Interaction of the bovine papillomavirus E6 protein with the clathrin adaptor complex AP-1. J Virol 72:476–482
49. Chen JJ, Reid CE, Band V, Androphy EJ (1995) Interaction of papillomavirus E6 oncoproteins with a putative calcium-binding protein. Science 269:529–531
50. Du M, Fan X, Hong E, Chen JJ (2002) Interaction of oncogenic papillomavirus E6 proteins with fibulin-1. Biochem Biophys Res Commun 296:962–969
51. Kiyono T, Hiraiwa A, Fujita M, Hayashi Y, Akiyama T, Ishibashi M (1997) Binding of high-risk human papillomavirus E6 oncoproteins to the human homologue of the Drosophila discs large tumor suppressor protein. Proc Natl Acad Sci USA 94:11612–11616
52. Nakagawa S, Huibregtse JM (2000) Human scribble (Vartul) is targeted for ubiquitin-mediated degradation by the high-risk papillomavirus E6 proteins and the E6AP ubiquitin-protein ligase. Mol Cell Biol 20:8244–8253
53. Glaunsinger BA, Lee SS, Thomas M, Banks L, Javier R (2000) Interactions of the PDZ-protein MAGI-1 with adenovirus E4-ORF1 and high-risk papillomavirus E6 oncoproteins. Oncogene 19:5270–5280
54. Lee SS, Glaunsinger B, Mantovani F, Banks L, Javier RT (2000) Multi-PDZ domain protein MUPP1 is a cellular target for both adenovirus E4-ORF1 and high-risk papillomavirus type 18 E6 oncoproteins. J Virol 74:9680–9693
55. Sedman SA, Barbosa MS, Vass WC, Hubbert NL, Haas JA, Lowy DR, Schiller JT (1991) The full-length E6 protein of human papillomavirus type 16 has transforming and transactivating activities and cooperates with E7 to immortalize keratinocytes in culture. J Virol 65:4860–4866
56. Morosov A, Phelps WC, Raychaudhuri P (1994) Activation of the c-fos gene by the HPV16 oncoproteins depends upon the cAMP-response element at -60. J Biol Chem 269:18434–18440
57. Dey A, Atcha IA, Bagchi S (1997) HPV16 E6 oncoprotein stimulates the transforming growth factor-beta 1 promoter in fibroblasts through a specific GC-rich sequence. Virology 228:190–199
58. Gewin L, Galloway DA (2001) E box-dependent activation of telomerase by human papillomavirus type 16 E6 does not require induction of c-myc. J Virol 75:7198–7201
59. Oh ST, Kyo S, Laimins LA (2001) Telomerase activation by human papillomavirus type 16 E6 protein: induction of human telomerase reverse transcriptase expression through Myc and GC-rich Sp1 binding sites. J Virol 75:5559–5566
60. Ristriani T, Masson M, Nominé Y, Laurent C, Lefevre JF, Weiss E, Travé G (2000) HPV oncoprotein E6 is a structure-dependent DNA-binding protein that recognizes four-way junctions. J Mol Biol 296:1189–1203
61. Ristriani T, Nominé Y, Masson M, Weiss E, Travé G (2001) Specific recognition of four-way DNA junctions by the C-terminal zinc-binding domain of HPV oncoprotein E6. J Mol Biol 305:729–739
62. Li X, Coffino P (1996) High-risk human papillomavirus E6 protein has two distinct binding sites within p53, of which only one determines degradation. J Virol 70:4509–4516
63. Thomas MC, Chiang C-M (2005) E6 oncoprotein represses p53-dependent gene activation via inhibition of protein acetylation independently of inducing p53 degradation. Mol Cell 17:251–264
64. Cole ST, Danos O (1987) Nucleotide sequence and comparative analysis of the human papillomavirus type 18 genome. Phylogeny of papillomaviruses and repeated structure of the E6 and E7 gene products. J Mol Biol 193:599–608
65. Pim D, Storey A, Thomas M, Massimi P, Banks L (1994) Mutational analysis of HPV-18 E6 identifies domains required for p53 degradation in vitro, abolition of p53 transactivation in vivo and immortalisation of primary BMK cells. Oncogene 9:1869–1876
66. Thomas M, Pim D, Banks L (1999) The role of the E6-p53 interaction in the molecular pathogenesis of HPV. Oncogene 18:7690–7700
67. Nominé Y, Ristriani T, Laurent C, Lefèvre JF, Weiss E, Travé G (2001) Formation of soluble inclusion bodies by hpv e6 oncoprotein fused to maltose-binding protein. Protein Expr Purif 23:22–32
68. Nominé Y, Charbonnier S, Ristriani T, Stier G, Masson M, Cavusoglu N, Van Dorsselaer A, Weiss E, Kieffer B, Travé G (2003) Domain substructure of HPV E6 oncoprotein: biophysical characterization of the E6 C-terminal DNA-binding domain. Biochemistry 42: 4909–4917

69. Zanier K, ould M'hamed ould Sidi A, Boulade-Ladame C, Rybin V, Chappelle A, Atkinson A, Kieffer B, Travé G (2012) Solution structure analysis of the HPV16 E6 oncoprotein reveals a self-association mechanism required for E6-mediated degradation of p53. Structure 20:604–617
70. Uversky VN, Roman A, Oldfield CJ, Dunker AK (2006) Protein intrinsic disorder and human papillomaviruses: increased amount of disorder in E6 and E7 oncoproteins from high risk HPVs. J Proteome Res 5:1829–1842
71. Xue B, Ganti K, Rabionet A, Banks L, Uversky VN (2013) Disordered interactome of human papillomavirus. Curr Pharm Des
72. Salmena L, Carracedo A, Pandolfi PP (2008) Tenets of PTEN tumor suppression. Cell 133:403–414
73. Maehama T, Dixon JE (1998) The tumor suppressor, PTEN/MMAC1, dephosphorylates the lipid second messenger, phosphatidylinositol 3,4,5-trisphosphate. J Biol Chem 273:13375–13378
74. Zhang XC, Piccini A, Myers MP, Van Aelst L, Tonks NK (2012) Functional analysis of the protein phosphatase activity of PTEN. Biochem J 444:457–464
75. Shen WH, Balajee AS, Wang J, Wu H, Eng C, Pandolfi PP, Yin Y (2007) Essential role for nuclear PTEN in maintaining chromosomal integrity. Cell 128:157–170
76. Waite KA, Eng C (2002) Protean PTEN: form and function. Am J Hum Genet 70:829–844
77. Podsypanina K, Ellenson LH, Nemes A, Gu J, Tamura M, Yamada KM, Cordon-Cardo C, Catoretti G, Fisher PE, Parsons R (1999) Mutation of Pten/Mmac1 in mice causes neoplasia in multiple organ systems. Proc Natl Acad Sci USA 96:1563–1568
78. Li J, Yen C, Liaw D, Podsypanina K, Bose S, Wang SI, Puc J, Miliaresis C, Rodgers L, McCombie R et al (1997) PTEN, a putative protein tyrosine phosphatase gene mutated in human brain, breast, and prostate cancer. Science 275:1943–1947
79. Fraser MM, Zhu X, Kwon C-H, Uhlmann EJ, Gutmann DH, Baker SJ (2004) Pten loss causes hypertrophy and increased proliferation of astrocytes in vivo. Cancer Res 64:7773–7779
80. Das S, Dixon JE, Cho W (2003) Membrane-binding and activation mechanism of PTEN. Proc Natl Acad Sci USA 100:7491–7496
81. Walker SM, Leslie NR, Perera NM, Batty IH, Downes CP (2004) The tumour-suppressor function of PTEN requires an N-terminal lipid-binding motif. Biochem J 379:301–307
82. Lee JO, Yang H, Georgescu MM, Di Cristofano A, Maehama T, Shi Y, Dixon JE, Pandolfi P, Pavletich NP (1999) Crystal structure of the PTEN tumor suppressor: implications for its phosphoinositide phosphatase activity and membrane association. Cell 99:323–334
83. Rahdar M, Inoue T, Meyer T, Zhang J, Vazquez F, Devreotes PN (2009) A phosphorylation-dependent intramolecular interaction regulates the membrane association and activity of the tumor suppressor PTEN. Proc Natl Acad Sci USA 106: 480–485
84. Vazquez F, Ramaswamy S, Nakamura N, Sellers WR (2000) Phosphorylation of the PTEN tail regulates protein stability and function. Mol Cell Biol 20:5010–5018
85. Ross AH, Gericke A (2009) Phosphorylation keeps PTEN phosphatase closed for business. Proc Natl Acad Sci USA 106:1297–1298

Chapter 12

Modeling Protein–Protein Complexes Using the HADDOCK Webserver "Modeling Protein Complexes with HADDOCK"

Gydo C.P. van Zundert and Alexandre M.J.J. Bonvin

Abstract

Protein–protein interactions lie at the heart of most cellular processes. Determining their high-resolution structures by experimental methods is a nontrivial task, which is why complementary computational approaches have been developed over the years. To gain structural and dynamical insight on an atomic scale in these interactions, computational modeling must often be complemented by low-resolution experimental information. For this purpose, we developed the user-friendly HADDOCK webserver, the interface to our biomolecular docking program, which can make use of a variety of low-resolution data to drive the docking process. In this chapter, we explain the use of the HADDOCK webserver based on the real-life Lys48-linked di-ubiquitin case, which led to the 2BGF PDB model. We demonstrate the use of chemical shift perturbation data in combination with residual dipolar couplings and further highlight a few other cases where our software was successfully used. The HADDOCK webserver is available to the science community for free at haddock.science.uu.nl/services/HADDOCK.

Key words Docking, Protein–protein interactions, Biomolecular complexes, NMR, Ubiquitinitation

1 Introduction

Protein–protein interactions are at the basis of cellular function. In order to understand and manipulate the cell and its processes, insight into biomolecular interactions at an atomic scale is required. Major genomic and proteomic initiatives are working toward this goal. However, while the size of the human proteome is predicted to be in the order of 20,000, the interactome, the network of all interacting proteins, is estimated to be more around 650,000 [1], with additional levels of complexity linked to the dynamics of the assemblies and the localization and time of expression of their components in the cell. To make things worse, crystallizing protein complexes has proven to be substantially more difficult than single chains; studying them by NMR is a nontrivial undertaking. To close the gap between the number of interactions and structural knowledge about them, computational approaches complementary to

Daisuke Kihara (ed.), *Protein Structure Prediction*, Methods in Molecular Biology, vol. 1137,
DOI 10.1007/978-1-4939-0366-5_12,

experimental methods have been devised. One of these is protein–protein docking which aims to predict the structure of a complex starting from atomic models of the unbound subunits.

Within the plethora of docking software available, one can distinguish between several classes. Most docking programs try to sample thoroughly the interaction space by generating conformations using computational methods such as correlation techniques (often FFT-based) or geometrical hashing, to name only a few. Their scoring functions are primarily based on shape and electrostatic complementarity [2]. Our docking software HADDOCK (*H*igh *A*mbiguity-*D*riven *DOCK*ing) [3, 4] takes a somewhat unique approach by being mainly data-driven, limiting the sampling to the regions of the interaction space defined by the data.

HADDOCK is capable of using ambiguous experimental data to drive the docking process, such as, NMR chemical shift perturbation (CSP) and mutagenesis data. It does this by transforming the data into ambiguous interaction restraints (AIRs) that define a large network of ambiguous distances between residues expected to be involved in the binding mode without imposing any specific orientation on the components. Since the original publication [3] HADDOCK has been extended to handle other NMR sources of information such as residual dipolar couplings (RDCs) [5] diffusion relaxation [6], and pseudo-contact shifts [7]. Other low-resolution data such as small angle X-ray scattering (SAXS) [8] and cross-link data from mass-spectrometry can also be used for scoring and/or generating models.

In addition to using ambiguous and low-resolution information, HADDOCK is also well known for its way of handling flexibility of the subunits and its final refinement in explicit solvent, which can either be water or, to represent a hydrophobic environment, DMSO [9]. Currently HADDOCK is the most cited software in its category [2] and more importantly, one of the best performing docking software based on the CAPRI (Critical Assessment of PRediction of Interactions) competition, a community-wide experiment that allows comparison of the success of docking software by doing blind tests [10, 11].

HADDOCK has been used to generate quite a number of models deposited in the PDB (~100 to date). One of those structures is the Lys48-linked di-ubiquitin (Ub2) complex (PDB: 2BGF [5]). As is well known, ubiquitin plays a major role in the ubiquitin proteasomal pathway, which is the main mechanism of protein degradation in eukaryotic cells. It is also involved in many regulatory pathways. The minimal required signal for flagging proteins to be degraded by the proteasome is Lys48-linked tetra-ubiquitin. To gain insight into the conformation of poly-ubiquitin in solution, Lys48-linked Ub2 has been investigated by NMR, which resulted in CSP, RDC, and ^{15}N-relaxation data [12].

In this chapter, we describe the steps to perform the docking of Lys48-linked Ub2 using HADDOCK, based on CSP and RDC data, mimicking the docking process that led to the 2BGF model. In the Materials section, we provide information about the HADDOCK web server and also describe the data files needed to run this protocol. In the Methods section, we first shortly describe the HADDOCK docking protocol, followed by a step-by-step tutorial describing how we tackled the Lys48-linked Ub2 case using the HADDOCK web server. In the Notes section, we provide more in-depth information about specific aspects of the docking process. We end this chapter with a Case Studies section illustrating some biologically relevant cases where HADDOCK was successfully used.

2 Materials

2.1 Software Requirements

The first requirement to run this protocol is of course access to the HADDOCK software. This can be either through a local copy running on a desktop computer or cluster or more conveniently using the HADDOCK web server, which will be used in this protocol. The HADDOCK software consists of a collection of Python and CNS (Crytallography and NMR System) [13, 14] scripts with additional tools written in various languages. CNS is used as the computational engine that performs the computationally intensive part such as the energy calculations, minimizations and molecular dynamics refinement stages, while the Python routines are used for controlling the data flow, scoring and performing various pre- and post-processing tasks.

To facilitate the use of the software, we have developed the user-friendly HADDOCK web server [15] accessible at haddock.science.uu.nl/services/HADDOCK. Besides eliminating possible dependencies, it also comes with additional error checking and other automatic procedures, which makes it more robust. A special version of the web server making use of European Grid Initiative (EGI, www.egi.eu) resources is also available via the WeNMR web site (www.wenmr.eu) [16]. To use our web server one first needs to register for a user-account. The user-accounts come in various flavors, each giving a different amount of control over the docking and its associated parameters:

- The *Easy interface,* which is usually sufficient and the most straightforward to use, allows the user to upload PDB structures and specify the active and passive residues that define the interface of each molecule.
- The *Expert interface* gives some more control over the docking. In addition to the features available to the Easy interface, it allows the user to manually define the histidine protonation states and to specify which residues should be treated as semi- or fully flexible (both steps are performed automatically at the Easy level).

Furthermore, it permits the upload of user-defined distance, dihedral and hydrogen-bond restraints files and fine-tuning of various restraining and sampling parameters, e.g., the number of structures generated at the various stages.

- Finally, the *Guru interface* allows the user to tweak every parameter as if one was running a local version of HADDOCK. This is also the interface that gives access to additional restraints such as RDCs and relaxation anisotropy data. Symmetry can also be imposed at this level. In addition, it gives full control over almost all parameters including the various force constants and scoring weights for the docking.

Next to these three main interfaces, the server provides four additional interfaces:

- A *Prediction interface*, similar in its input requirements to the Easy interface, but with settings tuned for the use of bioinformatics predictions for docking [17].
- A *Refinement interface*, which only performs the final refinement in explicit solvent for a binary complex (the provided structures should thus already be in proper orientation).
- A *Multibody interface*, which allows the simultaneous docking of up to six different molecules.
- A *File upload interface*, which allows a one click upload of a parameter file previously saved from the web server (useful to repeat a docking with slight changes in parameter settings for example), and a *tool to generate ambiguous distance restraints*, especially useful for multicomponent (>2) systems.

For this tutorial the user should have registered for access to the HADDOCK web server and requested guru access in order to be able to use the Guru interface.

2.2 Data Requirements

The main data requirements to perform a docking run are atomic structures of each of the subunits of the complex in PDB format. These should preferably be structures determined by X-ray crystallography or NMR spectroscopy, but homology models may also be used [18]. For this particular protocol, we use as starting structures the NMR-determined ubiquitin structure 1D3Z, which corresponds to an ensemble of 10 solution structures, and 1AAR, which is a crystal structure with two ubiquitin chains (thus in total an ensemble of 12 structures). The experimental data to drive the docking consist of distance restraints derived from NMR CSP data and orientational restraints derived from RDC data [12]. All necessary files can be found in the corresponding Extra Material at extras.springer.com from where you can download an archive containing the mentioned PDB files, already prepared for docking, the CSP and RDC data files and the HADDOCK/CNS restraint files derived from these data files.

3 Methods

In this section, we first shortly describe the docking protocol that HADDOCK uses so that the user gets a better insight into what happens during the docking process. This helps in understanding the parameters. In the second part, we discuss how to perform the Ub2 docking using CSP and RDC data using the web server. It is assumed that the user has downloaded the tutorial folder from the Extra Materials at extras.springer.com

3.1 The HADDOCK Docking Protocol

Docking in HADDOCK is performed in three consecutive stages:

1. *Rigid body docking* (it0): the subunits are placed randomly in space with an approximate spacing of 25Å between them and subjected to a rigid body energy minimization to form the complex.
2. *Semiflexible refinement* (it1): the top scoring models (default 200 out of 1,000) in the it0-stage are refined using a simulated annealing in torsion-angle space procedure during which the interface is treated as flexible (first side chains only, then both side-chains and backbone).
3. *Flexible refinement in explicit solvent* (itw): in this final stage, the models from it1 are subjected to a gentle restrained molecular dynamics simulation in an explicit solvent shell (either water or DMSO as membrane mimic).

For further details refer to refs. [3, 4].

3.2 Docking Lys48-Linked Ub2 Using the HADDOCK Web Server

In this section, we describe the process of setting up a docking run using the Guru interface (http://haddock.science.uu.nl/services/HADDOCK/haddockserver-guru.html) of the web server. In order to make sense of the docking parameters, some knowledge about the Lys48-linked Ub2 complex is useful. It consists of two ubiquitin subunits that are linked together by a Gly76-Lys48 isopeptide bond. The subunit with the linked Gly76 is called the Distal Domain (ubiD) and the subunit with the linked Lys48 the Proximal Domain (ubiP).

1. Open an Internet browser and go to haddock.science.uu.nl/services/HADDOCK. Choose the Guru interface. This opens up the docking input screen as displayed in Fig. 1. Sections can be expanded or folded by clicking with the left-mouse-button on the double arrows on the right of each section name.
2. First give a name to your docking run. No spaces or special characters other than "-" or "_" are allowed! We named the run di-ubiquitin_CSP_RDC.

home >> HADDOCK >>

HADDOCK

Software web portal

Home HADDOCK Whiscy CPORT DNA Publications HADDOCK Inc. Contact

WELCOME TO THE UTRECHT BIOMOLECULAR INTERACTION WEB PORTAL >>

This is the Guru interface to the HADDOCK docking program.
This interface provides full control over HADDOCK parameters, except multi-body docking, and supports a wide range of experimental restraints.
Unfold the menus by clicking on the double arrows. Submit your job by providing your username and password and press submit.

You may supply a name for your docking run (one word)
Name

First molecule

Second molecule

Distance restraints

If you specified that passive residues will be defined automatically, all surface residues will be selected within the following radius (in angstroms) around the active residues 6.5

Instead of specifying active and passive residues, you can supply a HADDOCK restraints TBL file (ambiguous restraints) Browse... No file selected.

You can supply a HADDOCK restraints TBL file with restraints that will always be enforced (unambiguous restraints) Browse... No file selected.

If one of your molecules is DNA/RNA, restraints are automatically created to preserve its structure. Uncheck this option if you are docking with unstructured DNA/RNA

Create DNA/RNA restraints?

HADDOCK deletes by default all hydrogens except those bonded to a polar atom (N, O). Uncheck this option if you have NOEs or other specific restraints to non-polar hydrogens

Remove non-polar hydrogens?

Random patches

Define randomly ambiguous interaction restraints from accessible residues

Center of mass restraints

Define center of mass restraints to enforce contact between the molecules

Force constant for center of mass contact restraints 1.0

Surface contact restraints

Define surface contact restraints to enforce contact between the molecules

Force constant for surface contact restraints 1.0

Random exclusion

Randomly exclude a fraction of the ambiguous restraints (AIRs)

Number of partitions for random exclusion (%excluded=100/number of partitions) 2.0

Sampling parameters

Parameters for clustering

Dihedral and hydrogen bond restraints

Noncrystallographic symmetry restraints

Symmetry restraints

Restraints energy constants

Residual dipolar couplings

Relaxation anisotropy restraints

Energy and interaction parameters

Scoring parameters

Advanced sampling parameters

Solvated docking parameters

Analysis parameters

Username and password
Username
Password

Submit Query

Home HADDOCK Whiscy CPORT DNA Publications HADDOCK Inc. Contact

2008 © NMR Department. All rights reserved. Webdesign by Marc van Dijk
XHTML | CSS

Fig. 1 Overview of the HADDOCK webserver Guru interface (accessible from http://haddock.science.uu.nl/services/HADDOCK). Each section can be expanded by clicking on the *double arrows* on the *right* of the various sections in *red*. In the current view, the distance restraints section is expanded, revealing the forms used to upload the user-defined distance restraints file and various control parameters related to this class of restraints (color figure online)

Table 1
Data used during the docking

Distal domain (ubiD)	
Active residues	8, 9, 46, 47, 48, 49, 51, 68, 72, 73
Passive residues	6, 10, 11, 12, 39, 52, 53, 54, 71, 74, 75, 76
RDCs	46 NH RDCs
Fully flexible segments	72–76
Proximal domain (ubiP)	
Active residues	8, 9, 47, 48, 51, 68, 70, 72, 73, 74, 76
Passive residues	6, 10, 11, 12, 39, 46, 49, 52, 53, 54, 71, 75
RDCs	46 NH RDCs
Fully flexible segments	48; 72–76
Intervector projection angle restraints (VEAN)	***Number of restraints***
Intermolecular	981
Intramolecular	972
Isopeptide bond (Gly76–Lys48)	*Unambiguous restraint distance (Å)*
O–NZ	2.25 ± 0.05
C–NZ	1.35 ± 0.05
C–CE	2.45 ± 0.05
CA–NZ	2.45 ± 0.05

3. Secondly we have to define the PDB file of the first molecule, ubiD, to be docked. Expand the section *First molecule*. At the entry *Where is the structure provided?* click on the dropdown menu next to it and select *I am submitting it*. Set *Which chain of the structure must be used?* to *A* (*see* **Note 1**). Next to *PDB structure to submit* press the *Browse...* button and move to the location where the tutorial data were unpacked. Go to the *pdbs/*directory and select the *1AAR_1D3Z_ensemble.pdb* file (*see* **Note 2**).
4. Specify the interface by defining active and passive residues (*see* **Note 3**). The residues that are considered active from the CSP data are listed in Table 1. Fill in the numbers of the active residues in the textbox next to *Active residues*. Since CSP data typically does not show all residues participating in the binding we also want to define passive residues. Fill in the residue numbers in the textbox next to *Define passive residues* as given in Table 1 (*see* **Note 4**).
5. Specify the *Segment ID to use during the docking* for the first molecule as A (*see* **Note 5**).
6. HADDOCK distinguishes between semi- and fully flexible segments (*see* **Note 6**). The semiflexible segments will be determined automatically during this docking run, but the fully flexible segments need to be defined manually. In this particular case, we want to give more freedom to residues involved in the isopeptide bond and also to the unstructured

C-terminus. Expand the *Fully flexible segments* subsection. Below *Segment 1* set *First number* to *72* and *Last number* to 76. This will make residues 72–76, to which the second ubiquitin is attached, fully flexible.

7. The C-terminus of ubiD should be uncharged, since Gly76 is covalently bound to ubiP. So uncheck *The C-terminus of your protein is negatively charged* box.
8. Expand the *Second molecule* section. Since we are dealing with a peptide-linked homodimer, the *Structure definition* of the second molecule, ubiP, will be the same as ubiD. Set these parameters identical to the values as in the *First molecule* section. However, another set of active and passive residues were derived for this chain, as displayed in Table 1. Again, fill in the active and passive residue numbers in their respective boxes. Set *Segment ID to use during the docking* to *B*. For this chain we define two segments to be fully flexible, namely residue 48, the lysine, and the unstructured C-terminal tail. Set *First number* and *Last number* for *Segment 1*–48 and for *Segment 2*–72 and 76, respectively.
9. In addition to the AIRs that HADDOCK generates using the active and passive residues, additional distance restraint files can be uploaded. In this particular case, we want to include the Gly76-Lys48 isopeptide bond as an unambiguous restraint. These restraints are predefined in the file *restraints/ubiD-ubiP_pepbond.tbl* and shown in Table 1 (*see* **Note 7**). To upload this file, unfold the *Distance restraints* section and click on the *Browse...* button next to *You can supply a HADDOCK restraints...(unambiguous restraints)*. Point to the file located on disk. Uncheck the *Randomly exclude a fraction of the ambiguous restraints (AIRs)* box (*see* **Note 8**).
10. Expand the section *Sampling parameters*. Increase the *Number of structures for rigid body docking* to 1,440 (we have 12 starting models for each ubiquitin, giving 144 combinations, each sampled ten times, which amounts to 1,440 models) (*see* **Note 9**). The other default sampling parameters do not need to be changed. So after the rigid body docking stage the top 200 scoring structures will be refined.
11. Go to the *Restraints energy constants* section and in the *Energy constants for unambiguous restraints* subsection change the entries *hot, cool1, cool2*, and *cool3* to 0.1, 1, 5, and 5, respectively.
12. Unfold the *Residual dipolar couplings* section. There are three RDC restraint files in the *restraints/*directory: two with Intervector Projection Angle (VEAN) restraints and the other with direct SANI restraints (*see* **Note 10**). Expand the *Residual dipolar couplings 1* subsection. Set *RDC type* to VEAN.

Expand the *SANI energy constants* subsection and set *First iteration* to 0 and *Last iteration* to 1. The file containing the intermolecular VEAN restraints is named *ubiDP_vean_inter.tbl* (*see* **Note 11**). Upload the file by clicking on the *Browse...* button and selecting it.

13. Expand the *Residual dipolar coupling 2* subsection to upload the second set of RDC restraints, which are intramolecular VEAN restraints. Set *RDC type* to *VEAN* again. In the *SANI energy constants* subsection change *First iteration* and *Last iteration* both to 1. Again, upload the restraints by clicking on *Browse...* and select the *restraints/ubiDP_vean_intra.tbl* file.
14. The third set of RDC restraints contains the previous two combined but as a SANI restraint. For this, set the *RDC type* to SANI and the *R* and *D* value to 0.057 and −11.49, respectively. The SANI restraints will only be used in the final refinement in explicit solvent (itw). Change the *First iteration* entry to 2 in the *SANI energy constants* subsection. The SANI restraint file is named *ubiDP_sani.tbl*, so define the *RDC file* accordingly.
15. We are now ready to send the docking run to the HADDOCK server. Fill in your Username and Password at the bottom of the screen and press the *Submit Query* button. This sends the information to the server and adds the docking run to the queue once properly validated. You should be redirected to a new page that allows you to download a HADDOCK parameter file containing all parameters and data for your docking run (it is recommended to save it—this file can be uploaded again to the File upload interface). The extra material contains an example of such a file. The page also gives a link to the page that shows the current status of the docking run and where the final results will appear. A confirmation message will be sent to the email address provided at registration.
16. After the docking run has completed, typically after a few hours depending on the server load, an email is sent informing you where to find the results (this is the same link as provided by the server page upon successful submission). By following the link in the email you will be redirected to the HADDOCK web server results page, as shown in Fig. 2.
17. The result page first displays the name of your docking run and its status. It provides you a link where you can download the complete docking run as a gzipped tar file for further manual analysis. Also the docking parameter file containing all of your input data and parameter settings can be downloaded.
18. After that a summary is given of the docking run, giving information about the number of clusters created and how many water-refined models do cluster in these. In this case you should see four clusters containing almost all water-refined models (*see* **Note 12**).

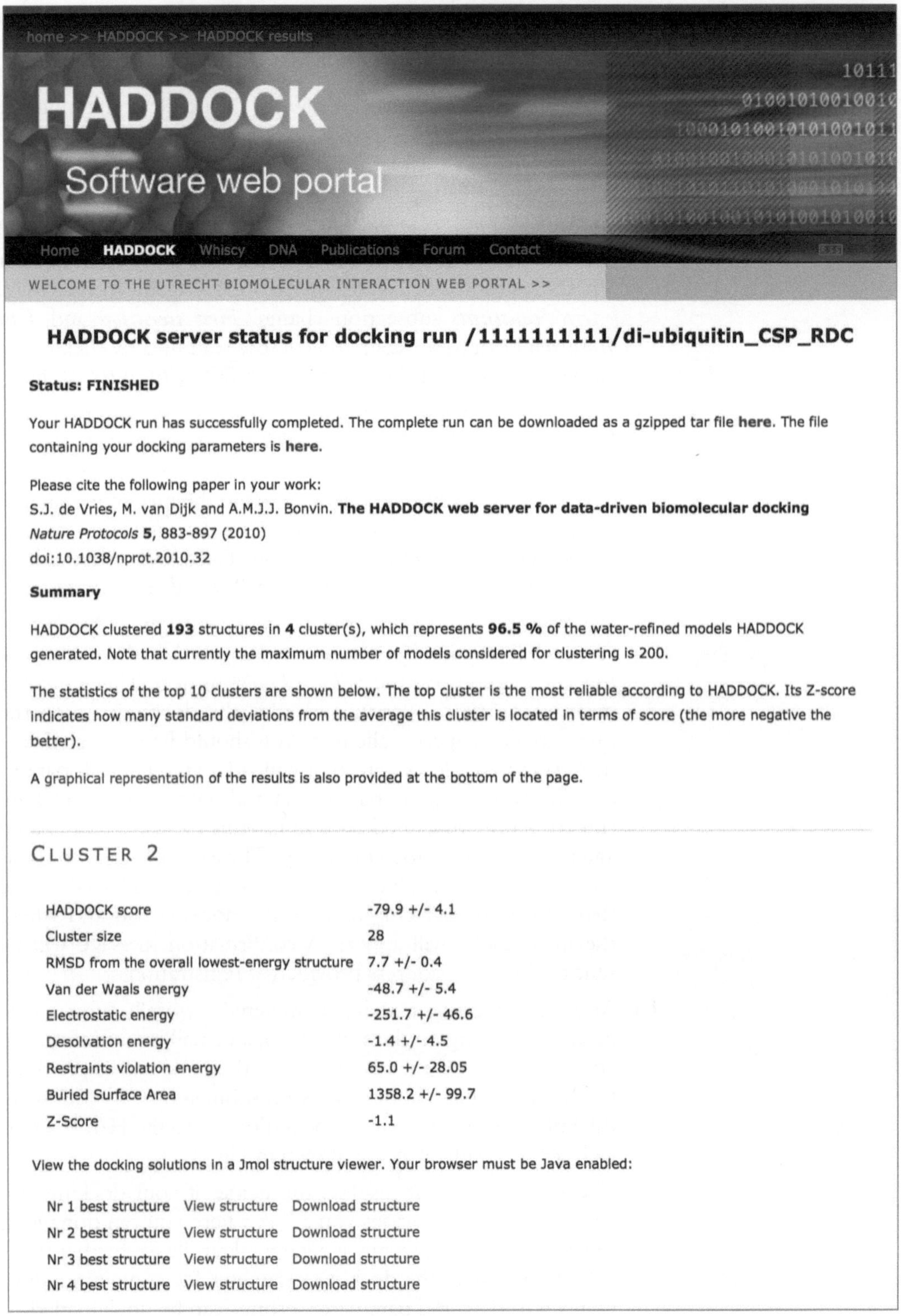

home >> HADDOCK >> HADDOCK results

HADDOCK

Software web portal

Home HADDOCK Whiscy DNA Publications Forum Contact

WELCOME TO THE UTRECHT BIOMOLECULAR INTERACTION WEB PORTAL >>

HADDOCK server status for docking run /1111111111/di-ubiquitin_CSP_RDC

Status: FINISHED

Your HADDOCK run has successfully completed. The complete run can be downloaded as a gzipped tar file **here**. The file containing your docking parameters is **here**.

Please cite the following paper in your work:
S.J. de Vries, M. van Dijk and A.M.J.J. Bonvin. **The HADDOCK web server for data-driven biomolecular docking**
Nature Protocols **5**, 883-897 (2010)
doi:10.1038/nprot.2010.32

Summary

HADDOCK clustered **193** structures in **4** cluster(s), which represents **96.5 %** of the water-refined models HADDOCK generated. Note that currently the maximum number of models considered for clustering is 200.

The statistics of the top 10 clusters are shown below. The top cluster is the most reliable according to HADDOCK. Its Z-score indicates how many standard deviations from the average this cluster is located in terms of score (the more negative the better).

A graphical representation of the results is also provided at the bottom of the page.

CLUSTER 2

HADDOCK score	-79.9 +/- 4.1
Cluster size	28
RMSD from the overall lowest-energy structure	7.7 +/- 0.4
Van der Waals energy	-48.7 +/- 5.4
Electrostatic energy	-251.7 +/- 46.6
Desolvation energy	-1.4 +/- 4.5
Restraints violation energy	65.0 +/- 28.05
Buried Surface Area	1358.2 +/- 99.7
Z-Score	-1.1

View the docking solutions in a Jmol structure viewer. Your browser must be Java enabled:

Nr 1 best structure	View structure	Download structure
Nr 2 best structure	View structure	Download structure
Nr 3 best structure	View structure	Download structure
Nr 4 best structure	View structure	Download structure

Fig. 2 Example view of a result page of the HADDOCK webserver. This view of the *top* part of the window shows the name of the docking run, its status, and gives information about the number of clusters found. Moreover, it provides detailed information on a per-cluster basis, with the values of the HADDOCK score and its various components indicated. In this view only the top scoring cluster is displayed

19. This is followed by a more rigorous analysis of each cluster. Only the ten best scoring clusters are shown maximally, so in this docking run all clusters are reported. The clusters are named sequentially based on their size, i.e., the largest cluster is named *Cluster 1*. However, the server returns the clusters in the order of their ranking based on the average HADDOCK score of the top four members of each cluster. The best scoring cluster, i.e., the cluster with the smallest HADDOCK score, is the first in the list. For this run *Cluster 2* appears at the top.
20. Each cluster section reports the various average scores with standard deviations based on the top four scoring structures of each cluster, as shown in Fig. 2. First the HADDOCK score (*see* **Note 13**) and the cluster size are given, then the RMSD value of the top four members of the cluster with respect to the overall lowest-energy structure. This is followed by the values of the individual energy terms used in the HADDOCK score, such as the van der Waals, electrostatic, desolvation and restraints violation energies, and ends with the buried surface area (BSA) in ångstrom (*see* **Note 14**) and the cluster *Z*-score (*see* **Note 15**). In addition, links are provided to the PDB files of the four best scoring structures in the cluster, which can be viewed online with Jmol or downloaded for further analysis.
21. After the individual cluster analysis, the results are displayed graphically in the *Results analysis* section, as displayed in Fig. 3. When you click on a plot a larger version appears in the browser. In each plot a dot represents a model and the color of the dot indicates the cluster to which it belongs. The cluster averages with standard deviations are displayed as colored triangles with associated error bars, based again on the top four scoring structures in the cluster. The first three plots show the HADDOCK score versus the interface-ligand-RMSD (i-l-RMSD), the i-RMSD, and the l-RMSD, respectively (*see* **Note 16**). The next plot displays the HADDOCK score versus the fraction of common contacts (FCC) (*see* **Note 17**). The last three plots show the van der Waals, electrostatics, and AIRs energy versus i-RMSD.
22. The web page ends with some supplementary information, which explains the abbreviations used and notifies you that the HADDOCK results will be deleted after a week. So make sure to download the docking run.
23. Congratulations, you performed and analyzed your first docking run using CSP and RDC data! To see whether everything worked out nicely you can compare the resulting structures to the published 2BGF PDB file, which can be downloaded from the PDB website (http://www.pdbe.org).

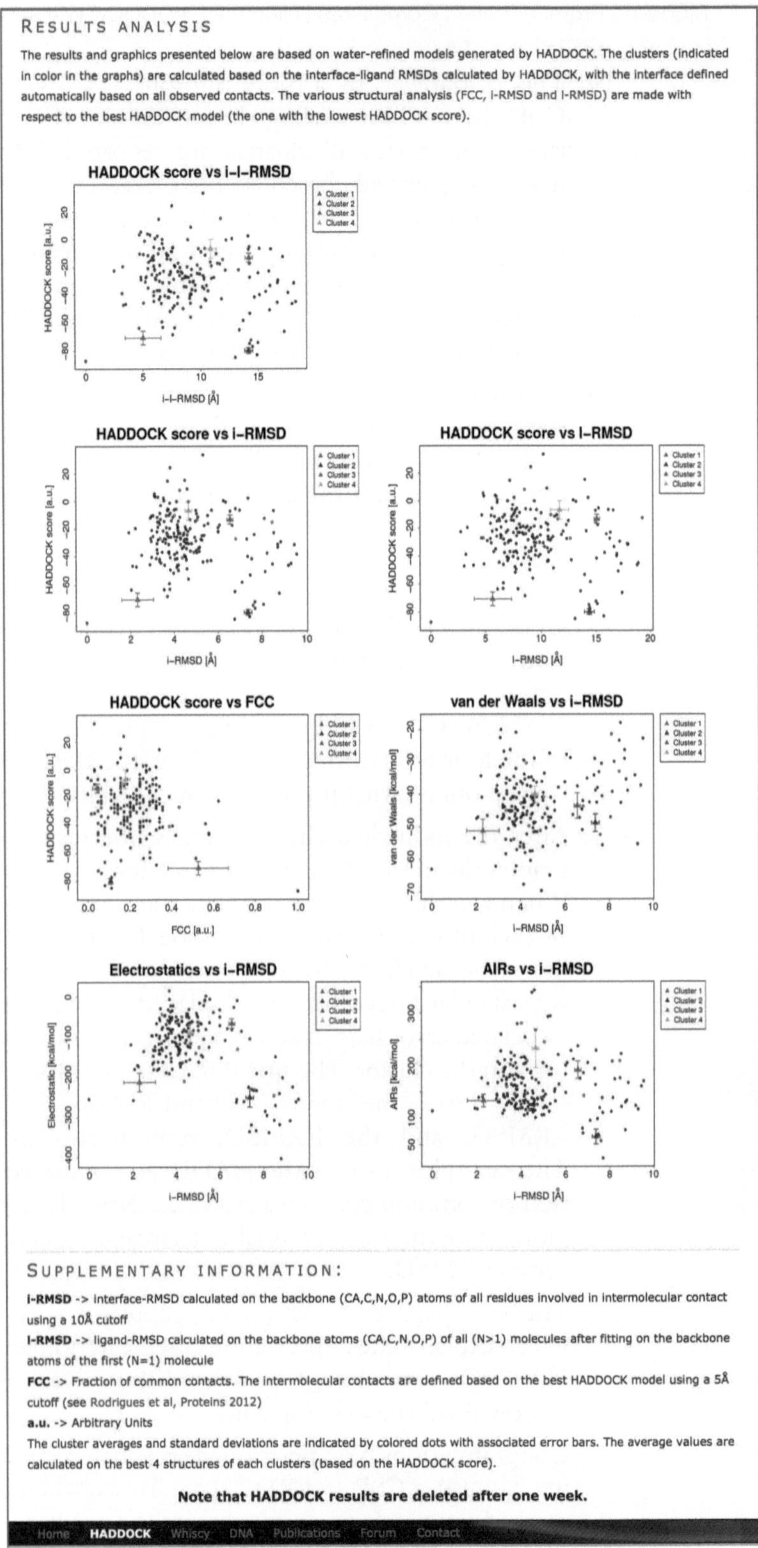

Fig. 3 Example view of a result page of the HADDOCK webserver. This view is of the *bottom* part of the window, which shows a graphical analysis of the results, displaying various scores and energy terms as well as cluster averages versus various RMSD values calculated with respect to the best scoring solution. The various clusters are color-coded. By clicking on a specific plot, an enlarged version is displayed for better viewing

4 Case Studies

The HADDOCK software has been used to solve quite a number of biologically relevant questions as illustrated by the high number of citations and the resulting models deposited in the PDB. One of these cases involved plectasin, a fungal defensin, and the bacterial cell-wall precursor Lipid II [22]. Defensins are host defense peptides that are part of the innate immune system and have antibiotic activity. Usually their activity is explained by their amphipathic structure, which binds and subsequently disrupts the microbial cytoplasmic membranes. Surprisingly, it was discovered that plectasin targets the bacterial cell-wall precursor Lipid II. HADDOCK was used, including CSP data, to unravel the primary binding mode of plectasin to Lipid II. The resulting model revealed that the interaction involved, via hydrogen bonding, the pyrophosphate moiety of Lipid II and several amide protons of plectasin. The resulting docking model, in combination with other experimental data, strongly supports a model in which plectasin gains affinity and specificity by binding to the solvent-exposed part of Lipid II, while its hydrophobic part interacts with the membrane (which was also revealed by NMR CSP data). Such studies can provide important insights for the development of new classes of antibiotics that are highly required considering the increase of resistant bacterial strains.

Other application examples of HADDOCK can be found in the latest CAPRI rounds [18, 23], where models of interactions have to be predicted in a blind manner. These are then compared to the actual crystal structure of the complex based on well-defined criteria. These criteria are i-RMSD, l-RMSD (as explained in **Note 16**), and the fraction of native contacts, which is the percentage of common residue contacts found in the binding mode of the predicted model with the crystal structure [11]. The models are ranked as either incorrect, acceptable, medium, or high quality. In the most recent CAPRI evaluation [23, 24], out of ten complexes that the HADDOCK group predicted, nine were of at least acceptable quality. This is a remarkable performance especially considering that several of the targets required first prediction of the structure of one of the components [18]. The modeling challenges were diverse and consisted of protein–protein complexes, dimers as well as multimers, a protein–polysaccharide complex, the prediction of the hydration structure at a protein–protein interface and even involved engineered interactions in designed complexes.

In conclusion, HADDOCK has gained a unique place among both the docking and experimental communities by being data-driven and having the abilities to handle flexibility and incorporate explicit water during the modeling process. Its user-friendly web interface makes it easily accessible to the science community. This is reflected by its large user group and the diverse scientific endeavors where it has been used to answer and provide insight into biology questions.

5 Notes

1. It is important to check the chainID values in the PDB file that you are uploading. If the chainID column is empty, simply set *Which chain of the structure must be used?* to *Any*. In this case, the ubiquitin models consist of only one chain denoted A.
2. The HADDOCK web server can deal with ensembles in PDB files as long as each model has the same number of residues and atoms. The PDB file *1AAR_1D3Z_ensemble.pdb* contains the 10 original NMR models from the 1D3Z entry and 2 from the 1AAR entry, making a total of 12 models. The models in the PDB file are separated by *MODEL/ENDMDL* statements. The two 1AAR chains were renumbered and processed to match the number of atoms and residues with 1D3Z and to make it compliant with the web server.
3. Active and passive residues are handled differently within the HADDOCK protocol as follows. HADDOCK generates distance restraints between the active residues of the first molecule and active and passive residues of the second molecule and vice versa. This means that, for each chain, active residues "feel" all active and passive residues of all other chains (unless specific chain selections are made using the *generate AIR* web server tool (http://haddock.science.uu.nl/services/GenTBL)). Contrarily, passive residues only "feel" active residues of other chains.
4. The active and passive residues were determined using 1H and 15N CSP data as follows. An active residue has a combined 1H and 15N CSP above average (0.033 ppm) and its backbone or side chain a relative solvent-accessible surface area of higher than 50 %. The solvent-accessible neighboring residues were defined as passive. Instead of manually selecting active residues, these can be automatically defined using the SAMPLEX software, which we developed (*see* ref. 19).
5. Make sure that the *Segment ID to use during the docking* is the same as was used during the creation of the restraint files. We gave ubiD the segment ID *A* and ubiP the segment ID *B* when creating the restraints.
6. The HADDOCK software distinguishes between semi- and fully flexible residues. Semiflexible residues become flexible during the last two stages of it1: First only their side-chain dihedral angles are allowed to vary, and then in the final simulated annealing stage both side chain and backbone are treated as flexible. Fully flexible residues are treated as flexible (both backbone and side-chain dihedral angles) from the start of the flexible refinement stage (it1), i.e., also during the high temperature searches.

7. The Gly76-Lys48 isopeptide bond is incorporated in this docking run as unambiguous distance restraints instead of treating it as a covalent bond (*see* Table 1). This allows the separation of the two ubiquitin chains at the beginning of the docking for a better sampling of conformations.
8. When *Randomly exclude a fraction of the ambiguous restraints (AIRs)* is checked (which is the case by default), a given percentage of restraints (default 50 %) is randomly discarded for each docking trial. In this way, "bad" data will be removed from time to time, ideally leading to better solutions. Thus, it provides a way to deal with false-positive predictions. When using bioinformatics predictions, this percentage can be as large as 87.5 % (default value on the Prediction interface server).
9. The entry *Number of structures for rigid body docking* defines the number of structures that are written to disk after the rigid body energy minimization. However, the parameter *Number of trials for rigid body energy minimization* gives the number of internal trials for the rigid body docking procedure. In addition, if the *Sample 180° rotated solutions during rigid body EM* box is checked then 180° rotated solutions with respect to the normal to the interface are automatically sampled. So, effectively, each model written to disk is the result of 10 docking trials (5 trials × 2 rotated solutions). For the di-ubiquitin case, the total number of models that are sampled amounts to 1,440 × 5 × 2 for a total of 14,400, or 100 docking poses per combination of starting structures (12 × 12).
10. The RDC restraints are incorporated into HADDOCK in two different ways. The first option is as direct SANI restraints where the two molecules are orientationally restrained with respect to an externally defined tensor. However, the RDCs can also be interpreted as Intervector Projection Angle Restraints [20] or VEAN restraints in CNS. This defines orientational restraints directly between two residues, eliminating the need of the cumbersome external tensor formalism, which has also been shown to facilitate the sampling in the initial structure calculation stages [20]. The use of VEAN restraints during the rigid body and first refinement stage and SANI restraints in the last refinement stage in water has been shown to be slightly superior in comparison to using only SANI or VEAN restraints [5].
11. Since the VEAN restraints represent intervector projection angles between two residues, the restraints can be divided into intermolecular and intramolecular restraints. During the rigid body energy minimization stage the intramolecular restraints serve no purpose since each molecule is kept rigid and so cannot change its conformation. Because of this, the intermolecular

restraints are used during the rigid body and first refinement stage, while the intramolecular restraints are only used in the first refinement stage.

12. Usually not all generated models do cluster, since a cluster should consist of minimally four structures maximally separated by an interface-ligand-RMSD (*see* **Note 16**) cut-off distance of 7.5 Å (by default). The minimal cluster size or cut-off value can be changed in the *Parameters for clustering* section at the Guru interface.
13. The HADDOCK score is a heuristic empirical function, which is a linear combination of several physical and empirical energy terms and a BSA term in the first stages. The HADDOCK score puts different weights on its components during each docking stage.
14. The BSA is calculated as follows. First the solvent-accessible surface area is calculated of each of the separated subunits and the modeled complex. The resulting BSA is the difference between the sum of the individual surface areas and the modeled complex.
15. The *Z*-score indicates how many standard deviations from the average a cluster is located in terms of its HADDOCK score. So the more negative the better.
16. All reported RMSDs are calculated with respect to the lowest scoring model (the best model according to the HADDOCK score). The i-l-RMSD, which is used for clustering if RMSD clustering is defined, is calculated on the interface backbone atoms of all chains except the first one after fitting on the backbone atom of the interface of the first molecule. The i-RMSD is calculated by fitting on the backbone atoms of all the residues involved in intermolecular contacts within a cutoff of 10 Å. The l-RMSD is obtained by first fitting on the backbone atoms of the first molecule and then calculating the RMSD on the backbone atoms of the remaining chains.
17. The FCC is the fraction of residue–residue contacts that two structures have in common. It can be used to cluster structures, and is faster, biologically more relevant, and has advantages in the case of symmetrical multimers in comparison to RMSD clustering [21].

Acknowledgments

Financial support from the Dutch Foundation for Scientific Research (NWO) (ECHO grant no. 711.011.009 and VICI grant no. 700.56.442) and the European Union (FP7 e-Infrastructure grant WeNMR no. 261572) is acknowledged.

References

1. Stumpf MPH, Thorne T, de Silva E et al (2008) Estimating the size of the human interactome. Proc Natl Acad Sci USA 105: 6959–6964
2. Moreira IS, Fernandes PA, Ramos MJ (2010) Protein-protein docking dealing with the unknown. J Comput Chem 31:317–342
3. Dominguez C, Boelens R, Bonvin AMJJ (2003) HADDOCK: a protein-protein docking approach based on biochemical or biophysical information. J Am Chem Soc 125: 1731–1737
4. de Vries SJ, van Dijk ADJ, Krzeminski M et al (2007) HADDOCK versus HADDOCK: new features and performance of HADDOCK2.0 on the CAPRI targets. Proteins 69:726–733
5. van Dijk ADJ, Fushman D, Bonvin AMJJ (2005) Various strategies of using residual dipolar couplings in NMR-driven protein docking: application to Lys48-linked di-ubiquitin and validation against 15N-relaxation data. Proteins 60:367–381
6. van Dijk ADJ, Kaptein R, Boelens R et al (2006) Combining NMR relaxation with chemical shift perturbation data to drive protein-protein docking. J Biomol NMR 34: 237–244
7. Schmitz C, Bonvin AMJJ (2011) Protein-protein HADDocking using exclusively pseudocontact shifts. J Biomol NMR 50:263–266
8. Karaca E, Bonvin AMJJ (2013) On the usefulness of ion-mobility mass spectrometry and SAXS data in scoring docking decoys. Acta Crystallogr D Biol Crystallogr 69:683–694
9. van Dijk ADJ, Bonvin AMJJ (2006) Solvated docking: introducing water into the modelling of biomolecular complexes. Bioinformatics 22: 2340–2347
10. Janin J (2005) Assessing predictions of protein-protein interaction: the CAPRI experiment. Protein Sci 14:278–283
11. Lensink MF, Wodak SJ (2013) Docking, scoring and affinity prediction in CAPRI. Proteins 81:2082–2095
12. Varadan R, Walker O, Pickart C et al (2002) Structural properties of polyubiquitin chains in solution. J Mol Biol 324:637–647
13. Brünger AT, Adams PD, Clore GM et al (1998) Crystallography & NMR system: a new software suite for macromolecular structure determination. Acta Crystallogr D Biol Crystallogr 54:905–921
14. Brünger AT (2007) Version 1.2 of the crystallography and NMR system. Nat Protoc 2:2728–2733
15. de Vries SJ, van Dijk M, Bonvin AMJJ (2010) The HADDOCK web server for data-driven biomolecular docking. Nat Protoc 5: 883–897
16. Wassenaar T, van Dijk ADJ, van Dijk M et al (2012) WeNMR: structural biology on the grid. J Grid Comp 10:743–767
17. de Vries SJ, Bonvin AMJJ (2011) CPORT: a consensus interface predictor and its performance in prediction-driven docking with HADDOCK. PLoS One 6:e17695
18. Rodrigues JPGLM, Melquiond ASJ, Karaca E et al (2013) Defining the limits of homology modelling in information-driven protein docking. Proteins 81:2119–2128
19. Krzeminski M, Loth K, Boelens R et al (2010) SAMPLEX: automatic mapping of perturbed and unperturbed regions of proteins and complexes. BMC Bioinformatics 11:51
20. Meiler J, Blomberg N, Nilges M et al (2000) A new approach for applying residual dipolar couplings as restraints in structure elucidation. J Biomol NMR 16:245–252
21. Rodrigues JPGLM, Trellet M, Schmitz C et al (2012) Clustering biomolecular complexes by residue contacts similarity. Proteins 80: 1810–1817
22. Schneider T, Kruse T, Wimmer R et al (2010) Plectasin, a fungal defensin, targets the bacterial cell wall precursor Lipid II. Science 328:1168–1172
23. Janin J (2013) The targets of CAPRI rounds 20-27. Proteins 81:2075–2081
24. Lensink MF, Méndez R, Wodak SJ (2007) Docking and scoring protein complexes: CAPRI 3rd edition. Proteins 69:704–718

Chapter 13

Predicting the Structure of Protein–Protein Complexes Using the SwarmDock Web Server

Mieczyslaw Torchala and Paul A. Bates

Abstract

Protein–protein interactions drive many of the biological functions of the cell. Any two proteins have the potential to interact; however, whether the interactions are of biological significance is dependent on a number of complicated factors. Thus, modelling the three-dimensional structure of protein–protein complexes is still considered to be a complex endeavor. Nevertheless, many experimentalists now wish to boost their knowledge of protein–protein interactions, well beyond complexes resolved experimentally, and for them to be able to do so it is important they are able to effectively and confidently predict protein–protein interactions. The main aim of this chapter is to acquaint the reader, particularly one from a non-computational background, how to use a state-of-the-art protein docking tool. In particular, we describe here the SwarmDock Server (SDS), a web service for the flexible modelling of protein–protein complexes; this server is freely available at: http://bmm.cancerresearchuk.org/~SwarmDock/. Supplementary files for Case Studies are provided with the chapter and available at extras.springer.com.

Key words SwarmDock, Protein–protein complexes, Protein–protein interactions, Protein docking, Protein structure prediction

1 Introduction

Protein–protein interactions are essential for the correct functioning of many biological processes, such as enzymatic catalysis and the immunological responsiveness of a cell to pathogens. However, for the protein interactome of essentially every species, including the human, experimental structures exist for only a small fraction of the possible functional complexes that are likely to form. Therefore, to boost studies on particular biological problems, effective use of theoretical tools for predicting the structure of protein–protein complexes is of considerable importance.

In general, modelling protein–protein interactions remains a complex problem, demanding careful design of powerful computational methods, from binding site prediction to generating and ranking (scoring) docked poses. Scoring docked poses, or potential

Daisuke Kihara (ed.), *Protein Structure Prediction*, Methods in Molecular Biology, vol. 1137,
DOI 10.1007/978-1-4939-0366-5_13,

solutions, is particularly important since a handful of correct solutions in a stack of a few thousand incorrect solutions may simply be missed, i.e., even if the particular docking algorithm is able to find correct solutions, the subsequent scoring scheme may not be able to place the correct solutions in a ranked list of top ten best solutions. A notable complicating factor for robust ranking of docked model poses is the number of conformational states that need to be searched, involving all coordinates for every atom forming the system. An experimental structure, as deposited in the Protein Data Bank (PDB; http://www.rcsb.org), is typically a single snapshot (X-ray structure) or a small ensemble of conformations (NMR structure) representing the system's full dynamic capacity. During the docking process, protein components making up a complex may change their conformational state only slightly, usually with movements restricted primarily to the protein side chains, and we term such cases to be examples of "rigid-body docking." On the other hand, more significant conformational changes could naturally occur or be induced for one or more of the docking partners, involving a range of both side chain and protein backbone movements, which we term "medium" or "difficult" docking cases depending on the predicted degree of conformational changes required for stable complex formation.

There are a number of different docking tools available, in the form of stand-alone programs or web services. Some of them are regularly tested in the blind trial for protein–protein docking, Critical Assessment of PRediction of Interactions (CAPRI [1]; http://www.ebi.ac.uk/msd-srv/capri/), where participants are typically given structures of receptor/ligand unbound molecules and asked to submit the best ten models for a complex that has been experimentally determined but for which the predictors have no knowledge. Predictors are expected to develop their own algorithms, which have usually been benchmarked against available experimental data, to test in these blind trials. An important database in this respect is provided by the Benchmark 4.0 dataset of protein–protein complexes where 176 structures of protein receptor/ligand pairs, in their unbound and bound forms are made available [2].

The quality of a docking tool is typically tested by docking unbound constituents and comparing each docked pose with the bound structure. Metrics used for such comparisons include calculating interface RMSD, ligand RMSD, and the fraction of native and nonnative contacts; this set of metrics enables classification of each solution to be in accordance with the CAPRI criteria, assigning the solution to be incorrect, acceptable, medium or of high quality [1].

In the last round of CAPRI (target T59, April/May 2013) the following publically available servers were registered for prediction:

1. pyDOCKWEB (http://life.bsc.es/servlet/pydock/home/),
2. GRAMM-X (http://vakser.bioinformatics.ku.edu/resources/gramm/grammx/),

3. DOCK/PIE (http://clsb.ices.utexas.edu/web/dock.html),
4. SwarmDock (http://bmm.cancerresearchuk.org/~SwarmDock/),
5. LZerD (http://kiharalab.org/proteindocking/lzerd.php),
6. SurFit (http://sysimm.ifrec.osaka-u.ac.jp/surFit/),
7. HADDOCK (http://haddock.chem.uu.nl),
8. ClusPro (http://cluspro.bu.edu).

The newly released SwarmDock Server (SDS) [3] participated in automatic prediction for the last four targets of the recently completed CAPRI 5th Assessment (2010–2013) and was the only server, indeed participant, that produced at least an acceptable prediction for each of the four targets (T53, T54, T57, T58). The core part of the server, the SwarmDock flexible protein–protein docking algorithm [4], was used in manual prediction mode for a number of other rounds, providing some highly competitive solutions. In addition, SDS has been carefully benchmarked on the complete Benchmark 4.0 test set and was able to find at least one correct solution for 122 complexes out of 176, with a correct solution in the best 10 (*see* Subheading 2) for 60 complexes out of 176 (CAPRI classification in ascending order of accuracy) [3]. The last value was recently significantly improved by using time-homogeneous finite state Markov chain models and by filtering out non-funnel-like structures [5]; an improvement which is currently being implemented into the online SDS.

In brief, SDS, is a docking tool which requires only a PDB formatted file for the unbound receptor and another for the unbound ligand as input, returning a ranked list of docked complex poses. The server has a variety of applications aimed at generally understanding, and even for the design of, protein–protein interactions. It may be used to: find the structure of an unknown protein–protein complex and in fact predict the unknown binding site (for blind docking benchmark see the Supplementary Information in ref. [3]); on a more theoretical note, structures generated may be used to sample the conformational states space and perform some analysis in terms of the theory of stochastic processes [5]; these structures, in comparison with the known complex, may be used to test potentials used for scoring [6]; finally it may be used to guide drug design studies to assess the binding of potential protein-based antagonists to a protein receptor of pharmaceutical importance.

2 Materials

The SDS may be accessed at: http://bmm.cancerresearchuk.org/~SwarmDock/.

The server's core is coded in C++, which is regularly maintained and updated. In order to create and maintain an interface between the user and our computational cluster, the web service is written in the modular language Python and uses the Common Gateway Interface.

The SwarmDock algorithm is based upon the principles of particle swarm optimization (PSO) [7] and has been described in detail elsewhere [3, 4, 8]. In brief, the binding energy between a receptor/ligand protein pair, as determined using the DComplex potential [9], is optimized using our own version of the PSO algorithm. To model the conformation of the receptor and the ligand, the position and orientation of the ligand, as well as normal mode coefficients, corresponding to the five lowest frequency nontrivial modes, make up the components of each particle search vector. After each PSO iteration, the lowest energy member undergoes a local optimization [10]. To ensure sufficient sampling of conformational space, the algorithm is run four times from each of approximately 120 starting positions, which are evenly spaced around the receptor.

To run a docking job PDB formatted files for both the designated receptor and ligand are required. There is no restriction on the number of chains for either receptor or ligand. Due to computer power limitations, the current number of atoms is restricted to 10,000 for both the receptor and ligand input files. However, if required, and upon request to the authors, users may be granted a higher limit. Docking may be performed in two modes: blind docking and restrained docking. The latter offers the opportunity to choose a list of the receptor's residues that are known (or are suspected) to form the interface of a specific protein–protein interaction and consequently restrict the search space.

Uploaded structures of the ligand and receptor must obey just three simple rules: files must have a TER record after each chain (also after the last one); only standard residues are allowed; there should be no missing residues or atoms (other than H). If there are any problems with the last two rules, the SDS attempts to fix them (*see* Subheading 3).

The PDB format is quite strict (http://www.wwpdb.org/docs.html); it restricts certain data types to specific columns. However, structures very often do not match it. The most prominent example is that atom types are in the wrong columns of the file (*see* Subheading 5 for details). During and after the submission process, the files are examined in several ways (*see* Subheading 3 for details). However, no one piece of software is currently able to predict all possible inconsistencies within a PDB submission file format. If the server returns an error, the users should feel free to contact the developer and ask for help if they are unsure as to the reason; any remaining, probably occurring with low frequency, formatting inconsistencies reported by users will be rectified at the computer code level.

As described here, SDS, in addition to its main docking function, can also act as a protein repairment service; uploaded structures of receptor and ligand molecules (each with one or several chains) in PDB format may be automatically repaired in several ways (*see* Subheading 3), which represents a unique feature in comparison to many other servers. We would like to point out that PDB files repaired by SwarmDock may be downloaded and used with other docking servers. In order to repair the files, the user needs to choose the restrained docking mode but does not need to resubmit the job for docking after repairment (*see* Subheading 3).

Computations may take up to a few days. Execution time depends not only on the size of the structure but also on the computational resources available at the time of submission. We may check the status of a job if requested by a user. There is currently no automatic time indicator on the Web site but this is in preparation. All docking results are deleted from the server 1 week after the user is emailed a message indicating job completion.

3 Methods

The main submission page for the SDS is depicted in Fig. 1 (upper part). The first step in using the server is to have two PDB formatted files stored somewhere on the user's computer, one for the designated ligand and one for the designated receptor. It is computationally more efficient (the server return time is shorter) for the user to select the larger protein as the receptor; however, either protein may be designated to be the receptor or ligand. Next the user is required to provide an email address and job name. Then by clicking the "Browse" buttons of the file-select fields, the PDB formatted files can be uploaded; the chosen file names will appear in the file upload box. The number of normal modes, for both the designated receptor and ligand, may be left as the default value. At this point certain checks are made by the server; see Subheading 5 for the details. The final step is to choose between the "Full blind" or "I want to choose interface residues" docking modes. If choosing the first mode (default), it is enough to click the "Submit new job" button and wait for an email with the link to the results page. After choosing the second mode, the button caption will change to "Choose residues." After clicking this, the job will be submitted only for preprocessing. Then the user will receive an email with the link to preprocessed receptor/ligand files (*see* further in this section) and a link to resubmit the job for docking. The resubmission form is depicted in Fig. 1 (lower part). There are two select boxes for the receptor and two for the ligand to choose a range of interface residues (Residue Type, Chain ID, and Residue Number). Please note that after preprocessing (in any mode) residues are numbered from 1, and chains are named from A to Z. The user can add any number of residues by iteratively choosing a residue range

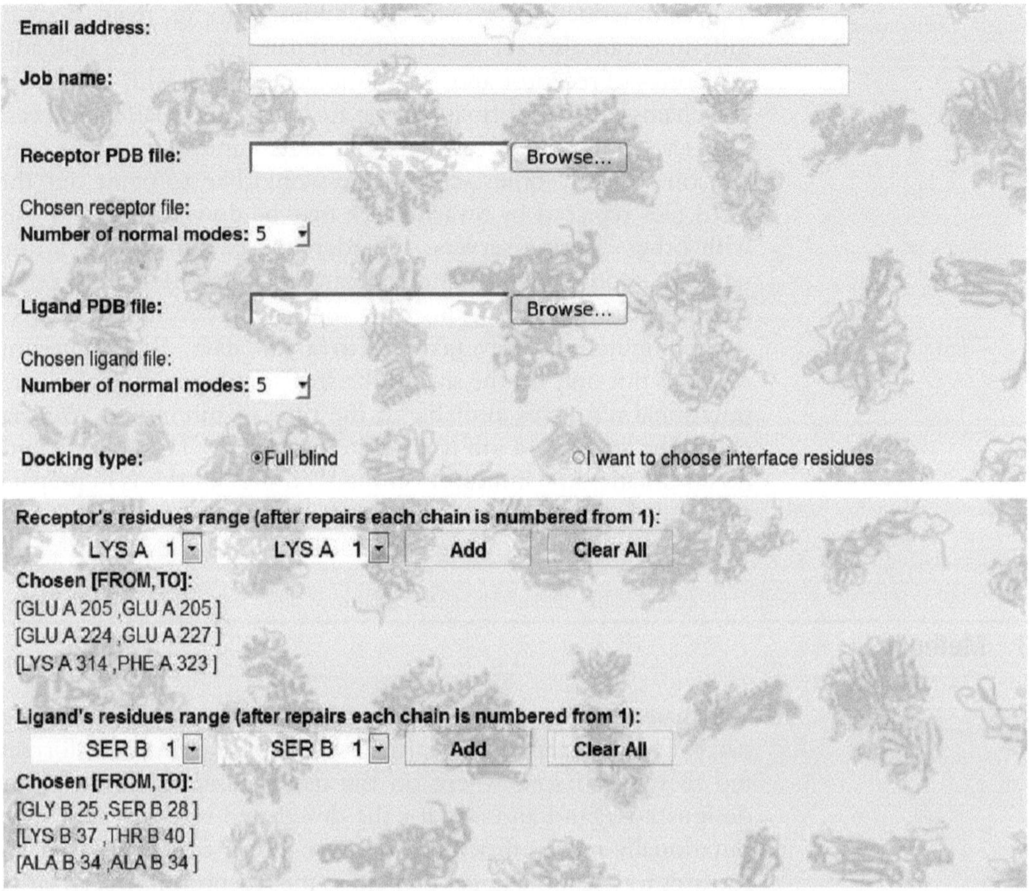

Fig. 1 The main (*upper panel*) part of the SDS submission page with the following elements: text fields for email address and job name, file-select fields for receptor and ligand PDB files, select boxes to choose number of normal modes for receptor and ligand (default value set to five), and radio buttons to choose the type of docking to be performed (blind or restrained, the former is the default option). After choosing the files, the chosen file name appears under the file-select fields. Restrained docking submission page (*lower panel*) with the following elements: two select boxes for a receptor and two for a ligand to choose a range of interface residues. A few restraints have already been chosen. The user can add as many residues as required by choosing residue ranges and clicking the "Add" button

and clicking the "Add" button. In the case of a mistake the user may click the "Clear All" button. As depicted, single residues as well as ranges may be chosen, additionally they may be chosen in any order. Residues chosen by the user will be copied to an output file called job.txt. The last step is to click the "Submit new job" button. After this, similar to the blind docking mode, the user needs to wait for an email informing them of job completion.

The server workflow consists of the following steps: preprocessing, docking, and postprocessing. The preprocessing stage includes checking for structural correctness, checking for iCodes, alternative atom locations, and modelling missing and nonstandard residues. Details of currently supported nonstandard residues may be found on the SDS webpage. Currently, other nonstandard residues or heteroatoms are ignored. However, if this leads to missing residues, gaps are modelled as a set of alanine residues. Submitting files with missing residues or missing atoms other than H is not encouraged. However, the server tries to repair files with missing residues by modelling loops with alanine with the aid of the program Loopy [11] and files with missing atoms by modelling missing side chains with the aid of the side chain replacement program SCWRL [12]. Missing backbone atoms cause the removal of a residue, which is remodelled as an alanine. After all repairs, input structures are minimized (50 steps of steepest descent, 100 steps of conjugate gradient, and 200 steps of adopted basis Newton–Raphson) using the CHARMM molecular mechanics package [13].

The first part of the docking stage employs approximately 120 starting positions, generated uniformly around the receptor, from which swarms of ligand conformations are subsequently docked. In the restrained docking mode, the user may choose the residues belonging to the binding site and consequently the server only accepts the starting positions which "see" (RayTracing) at least one of the chosen receptor's residues. Consequently, in the restrained mode the SDS is forced to provide docked poses on the receptor's surface on or near the region selected by the user. If ligand restraints have also been entered these will be combined with the receptor restraints to calculate the number of constrained contacts across each docked pose interface. This information may be used when analyzing clusters of docked solutions.

In the next phase normal modes for the receptor and ligand molecules are calculated using ElNemo [14]. As described in Subheading 2, the main part of the docking run consists of a hybrid PSO/local search. The complete docking process is repeated four times at each of approximately 120 starting positions. The search vector consists of the position and orientation of the ligand, as well as normal mode coefficients to model the conformation of both the receptor and ligand. Extensive benchmarking indicates that for most docking purposes the default setting of using five normal modes, for both the receptor and ligand, provides the better results; modes are ordered according to frequency, with the lower frequency modes selected first. However, varying the number of modes, up to the maximum value of 25, can provide even better results for certain targets. At present, we cannot offer definitive advice on the optimum number of modes to use for a particular target; however, if particularly large conformational changes are expected it is advisable to use a relatively low number of modes since only low frequency modes will be used in the algorithm, thereby absorbing the higher degree of conformational change. Conversely, if the protein is expected to exhibit only small conformational changes upon complex formation, essentially motion higher in frequency, a larger number of modes should be selected.

In the postprocessing stage, all structures are first minimized using the program CHARMM in a similar manner described above for the minimization of input structures. Hydrogen atoms are used during minimization but removed from the final structures. Subsequently and prior to clustering, solutions are scored using CP_TSC, a side chain centroid potential described by Tobi [15]. The clustering scheme first involves sorting the solutions in increasing value of CP_TSC. Then the lowest energy structure is selected as the first member of a new cluster. The RMSD between this structure and all other not yet clustered solutions is calculated, and solutions, if within a 3 Å threshold, are appended to the growing cluster. Subsequently the next lowest energy structure which had not yet been clustered is taken, and the cluster-growing scheme as described above repeated. This procedure of cluster seeding and growth is repeated until all solutions become members of a cluster. We observed that when a cluster contains a correct solution, that it is most often the lowest energy member of that cluster.

Upon completion of the computations, the user receives an email with the link to the results webpage. For successful termination, the webpage provides access to the docked complexes in the form of a compressed file containing a series of PDB formatted files. In addition all solutions and solutions forming the top ten-ranked clusters may be visualized using Jmol (Jmol plugin; http://jmol.sourceforge.net/). Here the geometrical center for each docked ligand is shown as a sphere, colored from blue to red in ascending order of interaction energy. In the case of restrained docking, the receptor's residues chosen by the user are shown in red.

In the compressed file, along with the PDB formatted structures for members of each cluster, a number of additional files are provided: clusters.txt (list of complexes within each ranked cluster); contacts.txt (list of contacts for the lowest energy member of each cluster); energies.txt (list of solutions with corresponding energies); best10.pdb (best 10 solutions in PDB format: the lowest energy structure from each of the first ten-ranked clusters); ligand.pdb and receptor.pdb (files used as input, may be different from those uploaded by the user due to repairs); uploaded_ligand.pdb and uploaded_receptor.pdb (files uploaded by the user), job.txt (details about the submitted job and list of restraints if applicable). Each output PDB file's name includes a number and a letter (a–d). The number refers to the starting position and the letter to the separate docking algorithm run from that particular starting position. These numbers can be thought of as a unique "label" for each solution.

After visual inspection, further analysis may be undertaken by interpreting the additional output files, where clusters.txt is the most important. Depicted in the upper part of Fig. 2 are the first ten lines from the cluster.txt file. Each line denotes a separate cluster and is constructed as follows: the lowest energy member of the cluster, its energy (energies for all structures may be found in the energies.txt file), number of members in the cluster, list of

```
Structure|Energy|Number of Members [Members]|Total Contacts|User Receptor's Residues Contacts|User Ligand's Residues Contacts
68c.pdb -28.57 1 [68c.pdb] 552 0 329
33b.pdb -26.79 3 [33b.pdb|32c.pdb|55a.pdb] 339 327 339
74a.pdb -24.56 5 [74a.pdb|61d.pdb|82a.pdb|86d.pdb|82c.pdb] 415 65 17
38d.pdb -24.30 115
[38d.pdb|6a.pdb|21a.pdb|8b.pdb|36c.pdb|13c.pdb|31a.pdb|12d.pdb|5c.pdb|4b.pdb|29c.pdb|13d.pdb|6c.pdb|28d.pdb|10b.pdb|30c.pdb|37c.
pdb|10c.pdb|18c.pdb|11d.pdb|32d.pdb|53a.pdb|15c.pdb|16a.pdb|11b.pdb|42d.pdb|9b.pdb|19a.pdb|50a.pdb|5d.pdb|54c.pdb|14a.pdb|39d.pd
b|25b.pdb|15d.pdb|16c.pdb|49d.pdb|41d.pdb|33a.pdb|31d.pdb|34c.pdb|12c.pdb|38c.pdb|21d.pdb|20c.pdb|14d.pdb|23a.pdb|8d.pdb|23c.pdb
|33c.pdb|22c.pdb|20d.pdb|12b.pdb|21c.pdb|1a.pdb|18b.pdb|29d.pdb|30d.pdb|49c.pdb|41b.pdb|24c.pdb|13a.pdb|2d.pdb|15a.pdb|14b.pdb|4
c.pdb|10a.pdb|17b.pdb|23d.pdb|7d.pdb|22d.pdb|1c.pdb|10d.pdb|34a.pdb|49b.pdb|9c.pdb|34b.pdb|25d.pdb|31b.pdb|5b.pdb|42a.pdb|26c.pd
b|38b.pdb|20b.pdb|22b.pdb|24b.pdb|32a.pdb|23b.pdb|22a.pdb|11c.pdb|25a.pdb|16b.pdb|35b.pdb|11a.pdb|35c.pdb|39b.pdb|14c.pdb|59b.pd
b|43c.pdb|5a.pdb|24d.pdb|40d.pdb|35a.pdb|8a.pdb|36a.pdb|31c.pdb|41a.pdb|33d.pdb|40b.pdb|39a.pdb|36b.pdb|45b.pdb|6d.pdb|15b.pdb|9
d.pdb] 349 343 349
73c.pdb -24.24 5 [73c.pdb|27d.pdb|68b.pdb|55c.pdb|46b.pdb] 469 6 102
76b.pdb -23.53 1 [76b.pdb] 439 125 5
40a.pdb -22.28 1 [40a.pdb] 371 357 350
98c.pdb -20.42 2 [98c.pdb|88c.pdb] 354 0 280
26a.pdb -18.75 1 [26a.pdb] 267 260 267
64c.pdb -18.68 8 [64c.pdb|54b.pdb|75b.pdb|43a.pdb|64a.pdb|53d.pdb|83b.pdb|54d.pdb] 294 247 144
```

```
Structure|Total Contacts|User Receptor's Residues Contacts|User Ligand's Residues Contacts|List of
Contacts(R-Receptor,L-Ligand,UR-User Receptor,UL-User Ligand)
68c.pdb 552 0 329
R:GLN A  30 :14;R:GLU A  79 :4;R:THR A  80 :12;R:TYR A  81 :11;R:GLU A  83 :13;R:GLU A  97 :9;R:LEU A 100 :13;R:GLN A 101
:42;R:HIS A 102 :12;R:ASP A 104 :7;R:ASP A 105 :58;R:ASN A 106 :6;R:PRO A 107 :3;R:ASN A 108 :11;R:TYR A 145 :47;R:ARG A 194
:22;R:CYS A 197 :6;R:ALA A 198 :2;R:LEU A 200 :14;R:GLN A 201 :58;R:LYS A 202 :19;R:THR A 240 :26;R:CYS A 242 :7;R:CYS A 243
:41;R:HIS A 244 :62;R:GLY A 245 :4;R:VAL A 459 :8;R:GLU A 462 :14;R:LYS A 463 :7;L:SER B   1 :10;L:HIS B   2 :33;L:MET B   3
:2;L:GLN B   7 :18;L:LEU B  10 :4;L:LYS B  11 :27;L:LYS B  14 :4;L:GLU B  15 :10;L:ILE B  18 :4;L:ASP B  29 :24;UL:PHE B  30
:71;UL:TYR B  31 :16;L:ASN B  33 :47;UL:ALA B  34 :24;L:ILE B  35 :6;L:ASN B  36 :34;UL:LYS B  37 :85;UL:ALA B  38 :12;UL:LYS
B  39 :43;UL:THR B  40 :6;UL:GLU B  42 :14;UL:GLU B  43 :9;UL:LEU B  47 :19;UL:GLU B  50 :20;UL:ILE B  51 :10;

33b.pdb 339 327 339
R:ARG A 206 :6;UR:LYS A 209 :11;UR:GLU A 224 :12;UR:PHE A 225 :4;UR:ALA A 226 :29;UR:GLU A 227 :11;UR:SER A 229 :22;UR:LYS A
230 :31;R:VAL A 232 :1;UR:THR A 233 :3;UR:TYR A 260 :8;UR:ASN A 264 :11;R:ASP A 266 :5;UR:SER A 267 :9;UR:ASN A 315 :23;UR:ALA
A 317 :1;UR:GLU A 318 :51;UR:ALA A 319 :30;UR:LYS A 320 :14;UR:ASP A 321 :9;UR:VAL A 322 :28;UR:PHE A 323 :9;UR:MET A 326
:11;UL:GLY B  25 :1;UL:ILE B  26 :16;UL:THR B  27 :12;UL:SER B  28 :14;UL:PHE B  30 :10;UL:TYR B  31 :38;UL:ALA B  34
:3;UL:LYS B  37 :15;UL:ALA B  38 :5;UL:LYS B  39 :21;UL:THR B  40 :2;UL:GLU B  42 :9;UL:GLU B  43 :43;UL:ALA B  46 :10;UL:LEU
B  47 :27;UL:GLU B  50 :28;UL:ILE B  51 :17;UL:LYS B  53 :7;UL:ALA B  54 :43;UL:HIS B  55 :18;

74a.pdb 415 65 17
R:GLN A 201 :5;R:LYS A 202 :20;R:PHE A 203 :61;R:GLY A 204 :7;R:ARG A 206 :32;R:ALA A 207 :29;R:ALA A 210 :3;R:TRP A 211
:9;R:CYS A 313 :7;UR:ALA A 317 :17;UR:GLU A 318 :5;UR:LYS A 320 :37;UR:ASP A 321 :6;R:LEU A 324 :2;R:LEU A 344 :26;R:LYS A 348
:8;R:GLU A 351 :11;R:GLU A 355 :24;R:LYS A 356 :1;R:CYS A 358 :32;R:ALA A 359 :23;R:SER A 477 :17;R:LEU A 478 :18;R:VAL A 479
:15;L:SER B   1 :80;L:HIS B   2 :50;L:MET B   3 :34;L:THR B   4 :33;L:ILE B   5 :33;L:GLN B   7 :1;L:TRP B   8 :9;L:LYS B  11
:34;L:GLU B  15 :13;L:ILE B  18 :4;L:LYS B  22 :9;L:ASP B  29 :31;UL:PHE B  30 :14;L:PHE B  32 :5;L:ASN B  33 :41;L:ASN B  36
:21;UL:LYS B  37 :3;
```

Fig. 2 Results from docking: (*upper panel*) first ten lines from the cluster.txt file (list of clusters); (*lower panel*) a few lines from the contacts.txt file (list of contacts). Visualization of the cluster.txt file is embedded into the results webpage. Energies for all structures may be found in the energies.txt file. *See* Subheading 3 for more details

members, total number of contacts (distance cutoff defined as the sum of van der Waals radii + 20 %) between receptor and ligand (excluding hydrogen atoms which are not present in the output structures), number of contacts from the receptor's residue list submitted by the user, number of contacts from the ligand's residue list submitted by the user; in the case of blind docking the last two numbers are set to 0. The lowest energy model complexes, in PDB format, from each of the top ten-ranked clusters are copied to best10.pdb. As a general rule, the user should take note not just of the lowest energy solutions but also of the size of each ranked cluster. If the first ranked cluster is of a relatively large size this is a good indication that it does contain a correct solution. However, for difficult docking cases, a cluster containing a single or a handful of solutions, further down the ranked list of clusters could also represent correct solutions. In the case of restrained docking, it is clearly of importance to take note of the number of contacts made between the chosen residues. Contacts information for the lowest energy solutions of each cluster is given in detail in the contacts.txt file. The first few lines from such a file are depicted in the lower part of

Fig. 2, where R denotes a receptor residue, L a ligand residue, UR a receptor residue chosen by the user, and UL a ligand residue. After residue type, chain ID, and residue number, the number of contacts is given. In the example presented in Fig. 2, in fact it is a complex with PDB code 2VDB described in more detail in Subheading 4 below, the lowest energy members have the following quality: the 33b.pdb is a high quality solution, 38d.pdb and 40a.pdb are medium quality, and 26a.pdb is of acceptable quality.

4 Case Studies

We describe here three docking targets that recently formed part of our benchmark set for the SDS server. They represent complexes that vary in their degree of difficulty to dock (*see* ref. 2) and consist of a different number of chains for either the designated receptor or ligand. All unbound structure input files were taken from the Benchmark 4.0 database [2] and checked for the presence of a TER record after each chain. In all examples we used the default setting of five normal modes for the receptor and five for the ligand. Files were submitted via the Web site as explained in Subheadings 2 and 3, in two modes, blind and restrained. Under restrained mode, interface residues IDs were taken from bound structures from the Benchmark 4.0 database. Solutions were evaluated by calculating interface RMSD, ligand RMSD, and the fraction of native and nonnative contacts that leads to the classification in accordance with the CAPRI criteria as incorrect, acceptable, medium or high quality, in ascending order of accuracy.

Supplementary Material to this chapter, available at extras.springer.com, contains the following directories for all complexes studied here: *BoundStructures* (split into receptor and ligand), *QualityOfSolutions* (in accordance to CAPRI rules), *Solutions* (SDS output files), *Visualization* (e.g., in order to obtain the visualization as in Fig. 3, use VMD, choose Orthographic display and BWR color scale, load the receptor file and depict as Purple/New Cartoon, load the ligands file and depict as Beta/VDW).

2VDB [19]: *protein–protein complex of serum albumin (3CX9) with peptostreptococcal albumin-binding protein GA module (2J5Y).*

Biological importance: bacterial surface protein domain binding to host serum protein.

Difficulty of docking: rigid-body.

Input files: one-chain receptor (3CX9, see uploaded_receptor.pdb, chain labelled A) and one-chain ligand (2J5Y, see uploaded_ligand.pdb, chain labelled A).

SDS repairs: removed first HIS residue from receptor input file due to missing N backbone atom, modelled 40 residues of the receptor

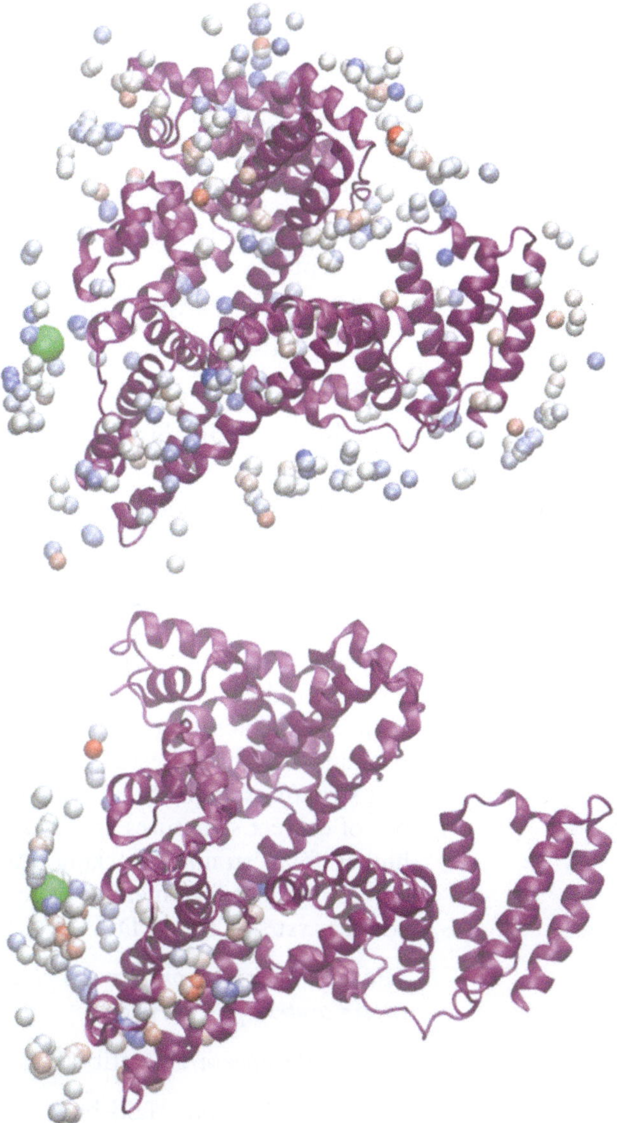

Fig. 3 The SDS solution space for 2VDB [19], the complex of serum albumin with peptostreptococcal albumin-binding protein GA module, from a blind (*upper figure*) and a restrained (*lower figure*) docking run. The geometrical center for each docked ligand pose is shown as a sphere, colored from *blue* to *red* in ascending order of interaction energy. Serum albumin is colored *purple*. The crystal conformation of the bound albumin-binding protein GA module is shown in *green*. Figures created using VMD [18]

with missing side chain atoms; renumbered residues in each chain starting from 1, receptor chain labelled as A, ligand chain as B; minimization; see receptor.pdb and ligand.pdb.

Residues chosen for restrained docking mode: chain A (205, 209, 224–227, 229–230, 233, 260, 264, 267, 305–306, 314–323, 326), chain B (25–28, 30–31, 34, 37–40, 42–43, 46–47, 50–51, 53–55); *see* receptor.pdb and ligand.pdb or job.txt file for details concerning residue type.

Number of output structures: 536 in blind docking, 400 in restrained docking.

Total number of correct solutions:

- Blind docking: 11 high quality, 15 medium quality, and 1 acceptable.
- Restrained docking: 1 high quality, 134 medium quality, and 10 acceptable.

Best10 results:

- Blind docking: high quality structure ranked as number 1 (21b.pdb)
- Restrained docking: high quality structure ranked as 2 (33b.pdb), medium quality structures ranked as 4 (38d.pdb) and 7 (40a.pdb), and acceptable structure ranked as 9 (26a.pdb).

In Fig. 3, we depicted the 2VDB solution space in blind (upper figure) and restrained (lower figure) mode. It can be seen that for restrained docking all solutions are close to the binding site. Figure 4 depicts correct solutions in the top10 obtained for restrained docking. Restrained docking helps to provide more correct structures in the best10. In addition, the ratio between number of correct structures and total number of output structures is higher for restrained docking. As mentioned previously, Fig. 2 depicts the first few lines of the restrained docking clusters.txt and contacts.txt files for 2VDB.

1E6J [20]: *protein–protein complex of Fab 13B5 (1E6O) with HIV-1 capsid protein p24 (1A43)*

Biological importance: antibody–antigen binding.

Difficulty of docking: rigid-body.

Input files: two-chains receptor (1E6O, see uploaded_receptor.pdb, chains labelled as L (light) and H (heavy) from antibody FAB fragment) and one ligand chain (1A43, see uploaded_ligand.pdb, chain unnamed).

SDS repairs: no repairs needed; renumbered residues of each chain starting from 1, receptor chains labelled as A and B, ligand chain as C; minimization; see receptor.pdb and ligand.pdb.

Residues chosen for restrained docking mode: chain A (90–92, 94), chain B (30–33, 50, 52–55, 57, 59, 99, 101–105), chain C (40, 54, 56–60, 62–63, 65–66, 69–70); see receptor.pdb and ligand.pdb or job.txt file for details concerning residue type.

Number of output structures: 422 in blind docking, 447 in restrained docking.

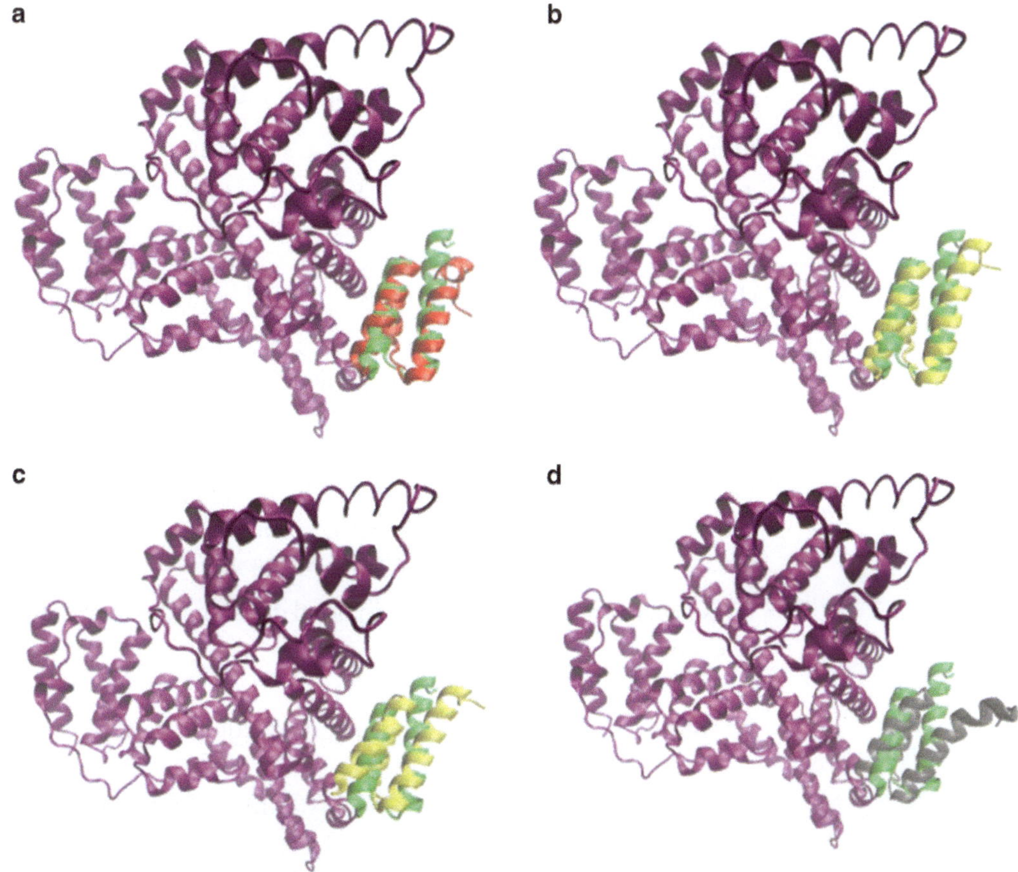

Fig. 4 Correct SDS solutions in the top10 for 2VDB [19] obtained by restrained mode docking and superimposed with Mustang [22] onto the bound complex. The bound receptor conformation of serum albumin is colored purple and the bound ligand conformation of albumin-binding protein GA module is shown in *green*; only the docked ligand conformations from the SDS solutions are shown, colored as follows: high quality in *red* (**a**—33b.pdb), medium quality in *yellow* (**b**—38d.pdb; **c**—40a.pdb), and acceptable in *gray* (**d**—26a.pdb). Figures created using VMD [18]

Total number of correct solutions:

- Blind docking: 0 high quality, 0 medium quality, and 1 acceptable.
- Restrained docking: 7 high quality, 1 medium quality, and 2 acceptable.

Best10 results:

- Blind docking: no correct structures in best10 (1a.pdb ranked as 35)
- Restrained docking: high quality structure ranked as 7 (19b.pdb).

Restrained docking helped to obtain a correct solution in the best10. In addition, with a comparable number of output structures, restrained docking provides more correct structures.

2HMI [21]: ***protein–protein complex of HIV1 reverse transcriptase (1S6P) with bound Fab 28 (2HMI)***

Biological importance: antibody–antigen binding.

Difficulty of docking: difficult.

Input files: two-chains receptor (1S6P, see uploaded_receptor.pdb, chains labelled as A and B) and two-chains ligand (2HMI, see uploaded_ligand.pdb, chains labelled as C (light chain) and D (heavy chain) of FAB fragment). Antibody is treated here as the ligand.

SDS repairs: no repairs needed; residues in each chain renumbered starting from 1, receptor chains labelled as A and B, ligand chains as C and D; minimization; see receptor.pdb and ligand.pdb.

Residues chosen for restrained docking mode: chain B (196, 199–200, 222–231, 358), chain C (32, 50, 91–92, 94, 96), chain D (30–33, 54–56, 60, 102–108); see receptor.pdb and ligand.pdb or job.txt file for details concerning residue type.

Number of output structures: 471 in blind docking, 284 in restrained docking.

Total number of correct solutions:

- Blind docking: 0 high quality, 0 medium quality, and 7 acceptable.
- Restrained docking: 0 high quality, 0 medium quality, and 10 acceptable.

Best10 results:

- Blind docking: no correct structures in the best10 (33c.pdb ranked as 16).
- Restrained docking: acceptable structure ranked as 10 (9c.pdb).

Restrained docking helped to achieve a correct solution in the best10. Figure 5 depicts correct solutions obtained for 1E6J in blind (1a.pdb) and restrained docking runs (19b.pdb), and those obtained for 2HMI in blind (33c.pdb) and restrained (9c.pdb) docking runs.

5 Notes

1. If there is no available structure for either the receptor or ligand, but the protein sequence is known and indicates a reasonable level of homology to an experimentally determined structure, the user may first build the structure(s) using a homology modelling server. A number of which are freely available (http://en.wikipedia.org/wiki/Homology_modeling), including several written and maintained in our own laboratory, 3D-JIGSAW [16] and POPULUS [17] (http://bmm.cancerresearchuk.org/services.html).

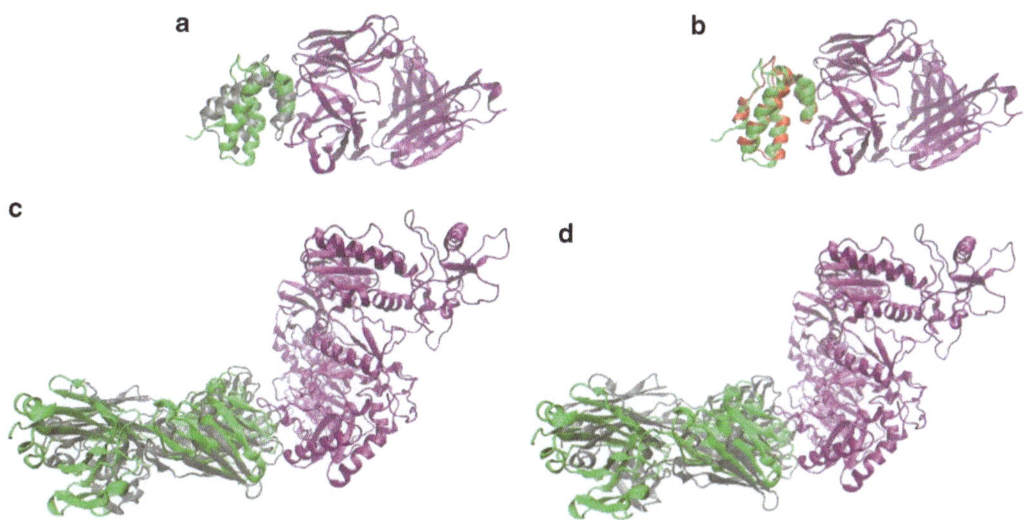

Fig. 5 Correct solutions obtained for 1E6J [20], complex of Fab 13B5 with HIV-1 capsid protein p24 in: (**a**) blind mode (1a.pdb), (**b**) restrained mode (19b.pdb). Correct solutions obtained for 2HMI [21], complex of HIV1 reverse transcriptase with bound Fab 28, in (**c**) blind mode (33c.pdb), (**d**) restrained mode (9c.pdb). The crystal conformation of each designated receptor (bound) is colored *purple* and designated ligand (bound) *green*; SDS solutions were superimposed with Mustang [22] onto the bound complex. High quality solutions are colored in *red* and acceptable in *gray*. Figure created using VMD [18]

2. The SDS algorithm is stochastic. This means that from identical submissions the user will obtain slightly different results. Consequently, the user may obtain a slightly better result from another submission. Moreover, results can be different according to which protein was designated the receptor and which the ligand. It is computationally more efficient for the larger protein to be designated the receptor. However, due to asymmetries in the starting density of points from which each swarm of ligands is set, experience has shown that on occasion a better set of docking solutions can be obtained by reversing the designation of the receptor/ligand pair.
3. Before submission the user should check that each chain is terminated with a TER record. Checks should also be made to see if all parts of each PDB file are in the correct columns. In case of any difficulties PDB files may be visualized, e.g., in VMD [18] and converted within the program into the correct PDB format that can be subsequently saved. Checks should also be made to ensure the PDB files have no formatting tags added by the text editor used or other strange characters.
4. SDS uses only the first 54 columns of the PDB file (up to the z coordinate).
5. The default number of normal modes equals five. Users can select up to 25. If there is no reason for using more normal modes, the default value should work fine.

6. The default number of stochastic swarm docking runs from each starting point equals four and, as benchmarked in the SDS application note [3], this leads to the best results. More runs may produce a higher number of correct solutions, but at the expense of generating more incorrect solutions. Options to use a higher density of starting positions, as well as more docking runs from each starting position, are available on request.
7. The submission is checked at the level of the web browser, as well as within the body of the SDS algorithm. Checks for which error messages are shown immediately are as follows: check to confirm if the email address is at least seven characters long and characters suit a regular expressions list; if the job name is at least three characters long and characters suit a regular expressions list; if the number of normal modes is in the range 1–25; if the receptor and ligand file were chosen and are not empty, if each ATOM or HETATM line is at least 54 characters long; if at least one TER record is present; if there is only one MODEL in the PDB file; if there is an ATOM record; if the PDB files were uploaded successfully to the server. Checks after the initial submission level that lead to special modes of action are as follows: if there are missing residues or side chain atoms, if unsupported nonstandard residues are used. After repairs (if needed), the structure is minimized. If the structure cannot be repaired or CHARMM is unable to minimize the structure, the user will receive an email with a link to an error webpage.
8. It is not always the case that a correct solution will be within the best ten, and not always will a correct solution be of high quality. Fully definitive protein–protein docking remains an unsolved problem. Therefore, usage of other protein docking servers is advisable; confidence will be boosted if a number of servers, which use different algorithms, return similar results. Structures may be firstly repaired with SDS and then used with alternative servers, a number of which are now freely available (http://en.wikipedia.org/wiki/Macromolecular_docking).
9. Finally, if the user experiences problems, they may write to us at: SwarmDock@cancer.org.uk; especially if an email has not been received within 1 week after job submission.

Acknowledgement

This work was funded by Cancer Research UK. We are grateful to Iain Moal and Raphael Chaleil for their fruitful collaborations on the subject of modelling protein–protein interactions.

References

1. Lensink MF, Mendez R, Wodak SJ (2007) Docking and scoring protein complexes: CAPRI 3rd edition. Proteins 69:704–718
2. Hwang H, Vreven T, Janin J, Weng ZP (2010) Protein-protein docking benchmark version 4.0. Proteins 78:3111–3114
3. Torchala M, Moal IH, Chaleil RAG, Fernandez-Recio J, Bates PA (2013) SwarmDock: a server for flexible protein-protein docking. Bioinformatics 29:807–809
4. Moal IH, Bates PA (2010) SwarmDock and the use of normal modes in flexible protein-protein docking. Int J Mol Sci 11:3623–3648
5. Torchala M, Moal IH, Chaleil RAG, Agius R, Bates PA (2013) A Markov-chain model description of binding funnels to enhance the ranking of docked solutions. Proteins 81:2143–2149. doi:10.1002/prot.24369
6. Moal IH, Torchala M, Bates PA, Fernandez-Recio J (2013) The scoring of poses in protein-protein docking: current capabilities and future directions. BMC Bioinformatics 14:286
7. Kennedy J, Eberhart RC (1995) Particle swarm optimization. In: Proceedings of the IEEE international conference on neural networks. Perth, Australia, pp 1942–1948
8. Li X, Moal IH, Bates PA (2010) Detection and refinement of encounter complexes for protein-protein docking: taking account of macromolecular crowding. Proteins 78:3189–3196
9. Liu S, Zhang C, Zhou H, Zhou Y (2004) A physical reference state unifies the structure-derived potential of mean force for protein folding and binding. Proteins 56:93–101
10. Solis FJ, Wets RJB (1981) Minimization by random search techniques. Math Oper Res 6:19–30
11. http://wiki.c2b2.columbia.edu/honiglab_public/index.php/Software:Loopy
12. Krivov GG, Shapovalov MV, Dunbrack RL Jr (2009) Improved prediction of protein side-chain conformations with SCWRL4. Proteins 77:778–795
13. Brooks BR, Brooks CL III, Mackerell AD Jr, Nilsson L, Petrella RJ, Roux B, Won Y, Archontis G, Bartels C, Boresch S, Caflisch A, Caves L, Cui Q, Dinner AR, Feig M, Fischer S, Gao J, Hodoscek M, Im W, Kuczera K, Lazaridis T, Ma J, Ovchinnikov V, Paci E, Pastor RW, Post CB, Pu JZ, Schaefer M, Tidor B, Venable RM, Woodcock HL, Wu X, Yang W, York DM, Karplus M (2009) CHARMM: the biomolecular simulation program. J Comput Chem 30:1545–1615
14. Suhre K, Sanejouand YH (2004) ElNemo: a normal mode web server for protein movement analysis and the generation of templates for molecular replacement. Nucleic Acids Res 32(Web Server issue):W610–W614
15. Tobi D (2010) Designing coarse grained-and atom based-potentials for protein-protein docking. BMC Struct Biol 10:40
16. Bates PA, Kelley LA, MacCallum RM, Sternberg MJ (2001) Enhancement of protein modeling by human intervention in applying the automatic programs 3D-JIGSAW and 3D-PSSM. Proteins Suppl 5:39–46
17. Offman MN, Tournier AL, Bates PA (2008) Alternating evolutionary pressure in a genetic algorithm facilitates protein model selection. BMC Struct Biol 8:34
18. Humphrey W, Dalke A, Schulten K (1996) VMD: visual molecular dynamics. J Mol Graph 14(33–38):27–28
19. Lejon S, Cramer JF, Nordberg P (2008) Structural basis for the binding of naproxen to human serum albumin in the presence of fatty acids and the GA module. Acta Crystallogr Sect F Struct Biol Cryst Commun 64:64–69
20. Monaco-Malbet S, Berthet-Colominas C, Novelli A, Battai N, Piga N, Cheynet V, Mallet F, Cusack S (2000) Mutual conformational adaptations in antigen and antibody upon complex formation between an Fab and HIV-1 capsid protein p24. Structure 8: 1069–1077
21. Ding J, Das K, Hsiou Y, Sarafianos SG, Clark AD Jr, Jacobo-Molina A, Tantillo C, Hughes SH, Arnold E (1998) Structure and functional implications of the polymerase active site region in a complex of HIV-1 RT with a double-stranded DNA template-primer and an antibody Fab fragment at 2.8 A resolution. J Mol Biol 284:1095–1111
22. Konagurthu AS, Whisstock JC, Stuckey PJ, Lesk AM (2006) MUSTANG: a multiple structural alignment algorithm. Proteins 64: 559–574

Chapter 14

DOCK/PIERR: Web Server for Structure Prediction of Protein–Protein Complexes

Shruthi Viswanath, D.V.S. Ravikant, and Ron Elber

Abstract

In protein docking we aim to find the structure of the complex formed when two proteins interact. Protein–protein interactions are crucial for cell function. Here we discuss the usage of DOCK/PIERR. In DOCK/PIERR, a uniformly discrete sampling of orientations of one protein with respect to the other, are scored, followed by clustering, refinement, and reranking of structures. The novelty of this method lies in the scoring functions used. These are obtained by examining hundreds of millions of correctly and incorrectly docked structures, using an algorithm based on mathematical programming, with provable convergence properties.

Key words Protein–protein docking, FFT-based docking, Knowledge-based potential, Atomic potential, Residue potential, Scoring function, Mathematical programming, Refinement and reranking

1 Introduction

The DOCK/PIERR protein docking server predicts the quaternary structure of the complex formed by two proteins, given their individual tertiary (3D) structures. The structures of the complexes can be useful in obtaining molecular details of protein function and biochemical pathways. Examples are interactions between an enzyme and its inhibitor or between an antibody and antigen. Further, given structural details of the interface between proteins, experiments can be designed to alter the strength and specificity of binding by introducing mutations at the interface. Finally, complexes can also aid in structure-based drug design, where designed small molecules can inhibit the interaction between two proteins by preferentially binding to one partner and thus affecting the pathways involving them [1, 2].

Protein–protein docking algorithms in general work in two stages. In the first stage, various possible conformations of the complex are examined and scored, treating the proteins as rigid bodies. The most frequently used methods for the search stage are

Daisuke Kihara (ed.), *Protein Structure Prediction*, Methods in Molecular Biology, vol. 1137, DOI 10.1007/978-1-4939-0366-5_14, © Springer Science+Business Media New York 2014

Fast Fourier Transforms [3–5], which enables fast exhaustive sampling of the search space, Monte-Carlo [6, 7] and Geometric Hashing [8]. In the second stage of refinement and reranking, some limited flexibility in the models is introduced through techniques like energy minimization [5, 9] and Monte-Carlo [7, 10], and structures are reranked with fine-grained scoring functions.

In DOCK/PIERR, the conformational space of complexes is sampled exhaustively using Fast Fourier Transforms, and the encountered structures are scored using a residue scoring function. This is followed by side-chain rearrangement of the proteins at the docking site and a short energy minimization. The structures are then rescored using a combination of residue and atomic scores. The novelty of this algorithm and its accuracy lies in the scoring functions used. These scoring functions are parameterized using mathematical programming [11] and provably optimal structural SVM algorithms [12]. Hundreds of millions of models encountered from docking hundreds of complexes are used in the learning, and the models include both correctly and incorrectly docked structures. Constraints that stipulate that the energy of a misdocked structure should be higher than the energy of a correctly docked structure are derived from these models. The set of constraints derived from all the models in the learning set is solved through methods like linear programming or structural SVMs, to produce the parameters of the scoring function. The docking algorithm has been tested on docking benchmark datasets and is found to perform comparable to the state-of-art docking algorithms [13], ranking fourth in the server category in the CAPRI assessment of 2013 [14].

2 Materials

2.1 Input

The server takes as input the PDB structures of the two proteins to predict the structure of the complex. *See* **Note 1** on details of how to prepare the PDB structures.

2.2 Program Description

One of the proteins (called receptor) is kept fixed. All possible rigid rotations and translations of the second protein (called ligand) with respect to the receptor are explored using Fast Fourier Transforms. Each conformation is scored using a linear combination of an interface residue-contact based scoring function, PIE, and a van der Waals-like term for shape complementarity. The top scoring models are then clustered and filtered for interface clashes. The top 1,000 models are then refined. The refinement involves side-chain remodeling of the interface residues using rotamers (SCWRL4 [15]) and a short rigid energy minimization in vacuum with the OPLS force field using the molecular dynamics package MOIL [16]. The last procedure removes bad contacts and makes the structures more chemically reasonable. The refined structures

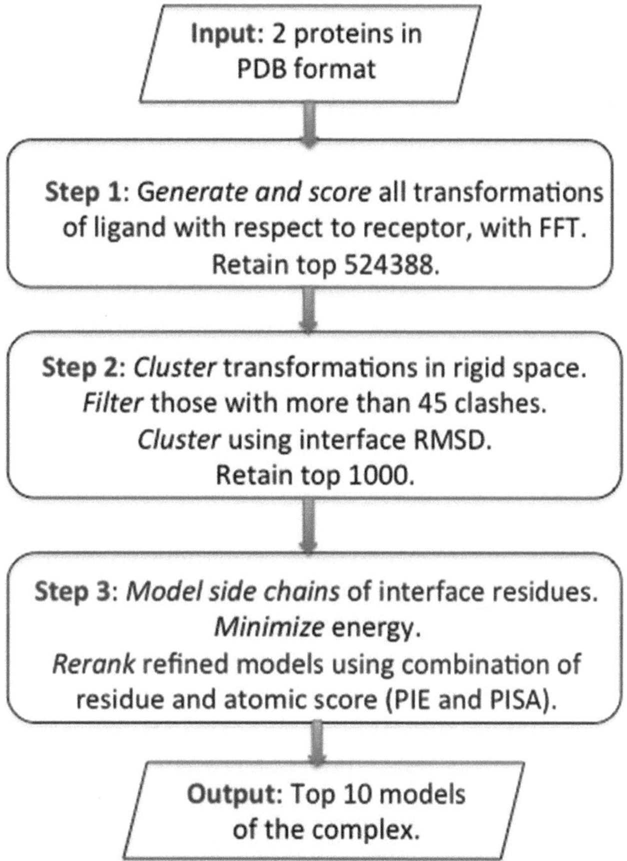

Fig. 1 Flowchart representing steps taken for docking two proteins using DOCK/PIERR

are then reranked using a combination of the residue potential, PIE, and an atomic potential, PISA, that was trained on refined models. It is to be noted that the adjustments during refinement are very small and typically of the order of ~0.1 A. Nevertheless they remove bad contacts and hence significantly improve the rescoring. The ten best ranked models of the complex are then made available as server output to the user. On tests on standard benchmarks and independent test sets, the algorithm as described above, obtains a near-native structure in the top ten models about 40–60 % of the time, and is comparable in accuracy to other leading docking algorithms. Figure 1 explains in detail the steps taken by the server to dock two proteins. For further details regarding the algorithm the reader is referred to [12, 13, 17].

2.3 Server Availability

The server is available at http://clsb.ices.utexas.edu/web/dock.html. It is implemented using HTML frontend and a PHP backend. The PHP script sends a mail to the server, which launches the docking jobs on 16 cores (Intel Xeon X5460, 3.16 GHz) of a Linux cluster at the University of Texas at Austin. While the entire docking package is in C++, the server also uses external programs such as SCWRL4 and MOIL.

2.4 Scoring Function Downloads

A user, who wishes to rank a set of structures obtained from a single server run or multiple-related runs, can also download and use our scoring functions, PIE (residue-based) and PISA (atomic). The source code and Linux executables for these are provided at http://clsb.ices.utexas.edu/web/dock_details.html. Scoring a model of a complex simply requires the structure of the complex in PDB format and the receptor and ligand chain names.

3 Methods

1. The server requires as input the structures of the two proteins in PDB format. The PDB files can be simply uploaded and submitted. *See* **Note 1** for potential sources of error in the input. Also *see* **Note 2** for cases where the user has only the sequence and not structure for an input protein.
2. For computational efficiency, the larger of the two proteins should be uploaded in the receptor field and the smaller one in the ligand field.
3. After submitting, the user gets a confirmation email with the job number. This job number denotes the submission ID and is referenced in the output email.
4. Jobs generally take about 4–5 h to complete. They may take more time if the proteins are large, i.e., longer than 400 residues, or if the server is experiencing high traffic.
5. Once the job is completed, a zipped file containing the ten best scoring docked conformations in PDB format is emailed to the user. The name of the zipped file corresponds to the submission ID or job number that the user was provided with, during submission. The chain names in the output PDB are alphabetically ordered, starting from the receptor chains.
6. Visualization of the models of the complexes can be performed with any structure visualization software like PyMol [18].
7. The accuracy of the docking method is between 40 % and 60 % currently, i.e., a near-native structure, a structure within 4 Å interface RMSD to the native, is in the top 10 docked structures about 40–60 % of the time. Cases where this docking method can be inaccurate are when the actual complex has a small number of contacts. Since (on the average) more contacts mean lower energy in our model, complexes with a small number of contacts are missed.

4 Case Studies

An early version of the docking software has been used previously in a biological study to suggest oligomeric conformations of a four-domain orange-fluorescent protein (Ember) [19]. Below we describe a case study of docking using DOCK/PIERR.

Unbound docking of Textilinin-1, a serine protease inhibitor with bovine trypsin. [*PDB 3D65*]

Here we dock bovine trypsin with the serine protease inhibitor, Textilinin-1, derived from the Australian Common Brown snake. This complex has been experimentally determined (PDB 3D65) [20]. Trypsin is an enzyme found in the pancreas and involved in proteolysis and digestion, while the protease inhibitor binds to trypsin to down-regulate its enzymatic activity.

To dock trypsin with its inhibitor, we perform unbound docking. That is, we model the tertiary structure of one or both of the constituent proteins using their homolog structures as templates. We then perform docking on the homology-modeled proteins. The trypsin molecule is chain E of the complex 3D65 and 223 residues long. The inhibitor molecule is chain I of 3D65 and 57 residues long. We use the structure of trypsin as in the bound form for docking, i.e., chain E of 3D65. To model the inhibitor, we perform a search for homologs using PSI-BLAST [21], searching the PDB database for structures homologous to chain I of PDB 3D65. We find that the chain I of PDB entry 3BTM is a good match, with *E*-value of 9×10^{-13} and sequence identity of 44.8 %. We next obtain pairwise alignments between the sequences of 3D65, chain I and 3BTM, chain I. A pairwise alignment can be obtained using dynamic programming, and is implemented in alignment servers such as the EMBOSS server (http://www.ebi.ac.uk/Tools/psa/). We then use the program Modeller [22, 23] to produce the structure of the inhibitor from the template structure of 3BTM, and the pairwise sequence alignment between 3BTM chain I and 3D65 chain I. We use the new PDB file obtained from Modeller for docking. Note that Modeller produces PDB files with no chain names by default, and hence it is recommended to add chain names to the PDB files before submitting files to the docking server.

We then submit the PDB files for the trypsin in the receptor field and the newly obtained inhibitor structure in the ligand field of the DOCK/PIERR server submission form. Upon completing the docking, we obtain the top ten models of the complex. Figure 2 shows the input proteins we docked, and Fig. 3 is a superposition of one of the top ten models obtained from the DOCK/PIERR server with the actual complex, 3D65. The model has an interface RMSD of 3.63 Å to 3D65.

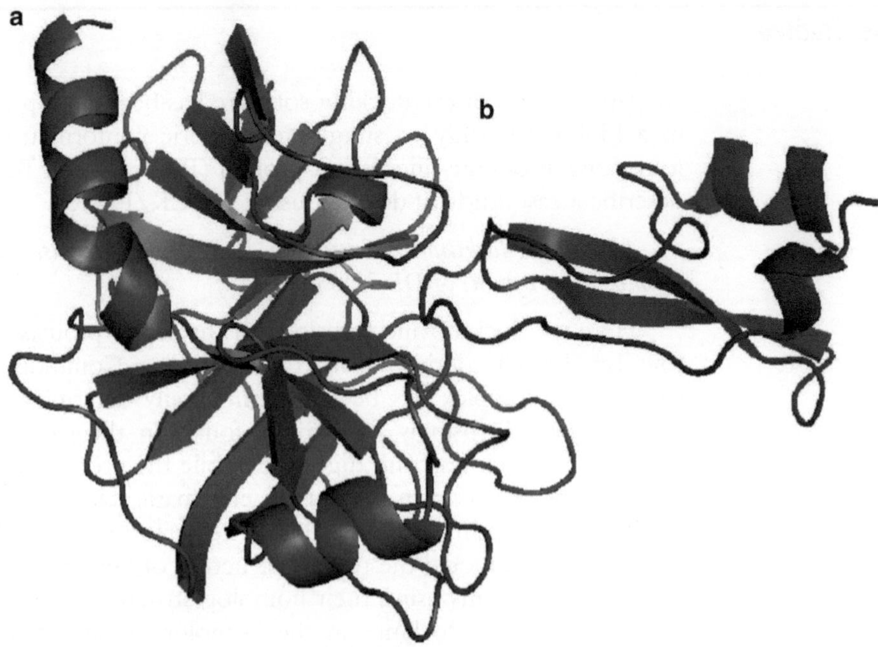

Fig. 2 (**a**) Chain E (bovine trypsin) and (**b**) Chain I (Textilinin-1, serine protease inhibitor) of complex 3D65 to be docked. These structures are inputs to the DOCK/PIERR server

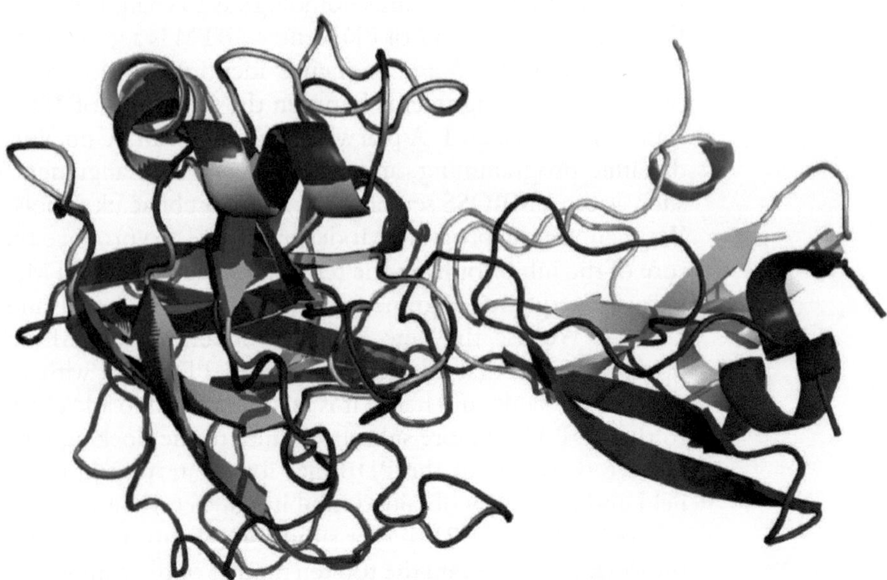

Fig. 3 One of the top ten models produced by DOCK/PIERR server (*cyan*) superposed with the native structure of the 3D65 trypsin–inhibitor complex (*blue*). The model has an interface RMSD of 3.63 Å to 3D65. The difference in tertiary structures between the native PDB and the model for the inhibitor is due to unbound docking (Color figure online)

5 Notes

1. The most common problems with the input PDB files that cause server failures are as follows:
   (a) Missing atoms in the PDB files. The missing atoms may be side-chain atoms or main chain atoms. For missing side-chain atoms, it is recommended to use the program SCWRL [15] or a similar program for side-chain placement. For missing main chain atoms, DOCK/PIERR is able to dock the proteins but the structures may not be refined, since the molecular dynamics program used in the refinement stage needs the coordinates of all the atoms. Failure to refine the models might result in less than optimal docking results.
   (b) Nonstandard atom names. These might be ignored in the initial docking stage and the structures may not be refined, as our molecular dynamics program is not capable of dealing with nonstandard atoms. These too might lead to suboptimal docking results if left unchanged.
   (c) Nonstandard residue names. Sometimes, some residues have nonstandard amino acid names. In many of these cases, the residue is chemically modified and the name is adjusted. For example the residue HIS is named differently as HSD, HSE, HSP depending on the protonation state. In such a case, the user is advised to rename such residues to their standard label.
   (d) Negative residue numbering. Some structures use negative residue numbers, for example when a tail is added to the native N-terminal. This causes problems during the refinement stage and the user is advised to index all residues with positive numbers.
   (e) Missing chain names for either protein, or identical chain names for both proteins. These can cause problems in the initial stages of docking. Also if the receptor and/or ligand have multiple chains, care must be taken to make all chain names between the receptor and ligand nonidentical. For example, if the receptor has chains A, B and the ligand has chain A, it is recommended to rename the ligand chain to C.
   (f) If a PDB file containing multiple NMR models is submitted, only the first model is considered for docking.
   (g) Some atoms in the PDB have multiple locations specified, using the alternate location field in the PDB. The docking program ignores the alternate locations. It also ignores HETATM records.

(h) Both the receptor and ligand molecules need to be proteins. Other molecules like DNA/RNA, or small molecule compounds are not supported as our scoring functions are tailored for protein interactions.

(i) If the user deals with very flexible peptides, it is recommended to dock one flexible conformation of the peptide at a time. This is because our algorithm performs rigid docking. At present it does not combine docking and internal motions.

2. It is most straightforward if one has the PDB structures of the two proteins to be docked. But if one just has the sequence for one (or both) input proteins, then structural model(s) need to be built from sequence. Examples of servers that produce models from the sequence include LOOPP [24, 25] (http://clsb.ices.utexas.edu/loopp/web/) and i-TASSER [26] (http://zhanglab.ccmb.med.umich.edu/I-TASSER/). If one has already a template structure on which to model the sequence, homology modeling packages such as Modeller [22, 23] (http://salilab.org/modeller) can be used to predict the structure. Note that using different templates or different modeling methods for structure prediction can affect docking results.

Acknowledgements

The authors acknowledge funding from NIH grant GM59796 and Welch grant F-1783.

References

1. Gray JJ (2006) High-resolution protein-protein docking. Curr Opin Struct Biol 16(2):183–193. doi:10.1016/J.Sbi.2006.03.003
2. Janin J, Bahadur RP, Chakrabarti P (2008) Protein-protein interaction and quaternary structure. Q Rev Biophys 41(2):133–180. doi:10.1017/S0033583508004708
3. Chen R, Li L, Weng ZP (2003) ZDOCK: an initial-stage protein-docking algorithm. Proteins Struct Funct Genetics 52(1):80–87. doi:10.1002/Prot.10389
4. Comeau SR, Gatchell DW, Vajda S, Camacho CJ (2004) ClusPro: a fully automated algorithm for protein-protein docking. Nucleic Acids Res 32:W96–W99. doi:10.1093/Nar/Gkh354
5. Tovchigrechko A, Vakser IA (2005) Development and testing of an automated approach to protein docking. Proteins 60(2):296–301. doi:10.1002/Prot.20573
6. Gray JJ, Moughon S, Wang C, Schueler-Furman O, Kuhlman B, Rohl CA, Baker D (2003) Protein-protein docking with simultaneous optimization of rigid-body displacement and side-chain conformations. J Mol Biol 331(1):281–299. doi:10.1016/S0022-2836(03)00670-3
7. Wang C, Bradley P, Baker D (2007) Protein-protein docking with backbone flexibility. J Mol Biol 373(2):503–519. doi:10.1016/J.Jmb.2007.07.050
8. Duhovny D, Nussinov R, Wolfson HJ (2002) Efficient unbound docking of rigid molecules. Lect Notes Comput Sci 2452:185–200
9. Li L, Chen R, Weng ZP (2003) RDOCK: refinement of rigid-body protein docking predictions. Proteins Struct Funct Genetics 53(3):693–707. doi:10.1002/Prot.10460

10. Wang C, Schueler-Furman O, Baker D (2005) Improved side-chain modeling for protein-protein docking. Protein Sci 14(5): 1328–1339. doi:10.1110/Ps.041222905
11. Wagner M, Meller J, Elber R (2004) Large-scale linear programming techniques for the design of protein folding potentials. Math Program 101(2):301–318. doi:10.1007/S10107-004-0526-7
12. Ravikant DVS, Elber R (2011) Energy design for protein-protein interactions. J Chem Phys 135(6):065102. doi:10.1063/1.3615722
13. Viswanath S, Ravikant DVS, Elber R (2013) Improving ranking of models for protein complexes with side chain modeling and atomic potentials. Proteins 81(4):592–606. doi:10.1002/prot.24214
14. Lensink M, Wodak SJ (2013) Docking, Scoring and Affinity Prediction in CAPRI. 81(12):2082–2095. doi: 10.1002/prot.24428
15. Krivov GG, Shapovalov MV, Dunbrack RL (2009) Improved prediction of protein side-chain conformations with SCWRL4. Proteins 77(4):778–795. doi:10.1002/Prot.22488
16. Elber R, Roitberg A, Simmerling C, Goldstein R, Li HY, Verkhivker G, Keasar C, Zhang J, Ulitsky A (1995) Moil—a program for simulations of macromolecules. Comput Phys Commun 91(1–3):159–189
17. Ravikant DVS, Elber R (2010) PIE-efficient filters and coarse grained potentials for unbound protein-protein docking. Proteins 78(2):400–419. doi:10.1002/Prot.22550
18. The PyMOL Molecular Graphics System, Version 1.5.0.4 Schrödinger, LLC.
19. Hunt ME, Modi CK, Aglyamova GV, Ravikant DVS, Meyer E, Matz MV (2012) Multi-domain GFP-like proteins from two species of marine hydrozoans. Photochem Photobiol Sci 11(4):637–644. doi:10.1039/c1pp05238a
20. Millers E-KI, Lavin MF, de Jersey J, Masci PP, Guddat LW. Crystal structure of textilinin-1, a Kunitz-type serine protease inhibitor from the Australian Common Brown snake venom, in complex with trypsin. *RCSB PDB entry 3D65*, http://www.rcsb.org/rxplore/explore.do?structureId=3d65. Accessed 5 June 2013
21. Altschul SF, Madden TL, Schaffer AA, Zhang J, Zhang Z, Miller W, Lipman DJ (1997) Gapped BLAST and PSI-BLAST: a new generation of protein database search programs. Nucleic Acids Res 25(17):3389–3402
22. Sali A, Blundell TL (1993) Comparative protein modeling by satisfaction of spatial restraints. J Mol Biol 234(3):779–815
23. Eswar N, John B, Mirkovic N, Fiser A, Ilyin VA, Pieper U, Stuart AC, Marti-Renom MA, Madhusudhan MS, Yerkovich B, Sali A (2003) Tools for comparative protein structure modeling and analysis. Nucleic Acids Res 31(13):3375–3380. doi:10.1093/Nar/Gkg543
24. Vallat BK, Pillardy J, Majek P, Meller J, Blom T, Cao B, Elber R (2009) Building and assessing atomic models of proteins from structural templates: learning and benchmarks. Proteins 76(4):930–945. doi:10.1002/prot.22401
25. Vallat BK, Pillardy J, Elber R (2008) A template-finding algorithm and a comprehensive benchmark for homology modeling of proteins. Proteins 72(3):910–928. doi:10.1002/prot.21976
26. Roy A, Kucukural A, Zhang Y (2010) I-TASSER: a unified platform for automated protein structure and function prediction. Nat Protoc 5(4): 725–738. doi:10.1038/nprot.2010.5

Chapter 15

Pairwise and Multimeric Protein–Protein Docking Using the LZerD Program Suite

Juan Esquivel-Rodriguez, Vianney Filos-Gonzalez, Bin Li, and Daisuke Kihara

Abstract

Physical interactions between proteins are involved in many important cell functions and are key for understanding the mechanisms of biological processes. Protein–protein docking programs provide a means to computationally construct three-dimensional (3D) models of a protein complex structure from its component protein units. A protein docking program takes two or more individual 3D protein structures, which are either experimentally solved or computationally modeled, and outputs a series of probable complex structures.

In this chapter we present the LZerD protein docking suite, which includes programs for pairwise docking, LZerD and PI-LZerD, and multiple protein docking, Multi-LZerD, developed by our group. PI-LZerD takes protein docking interface residues as additional input information. The methods use a combination of shape-based protein surface features as well as physics-based scoring terms to generate protein complex models. The programs are provided as stand-alone programs and can be downloaded from http://kiharalab.org/proteindocking.

Key words Protein–protein docking, Multiple-protein docking, Multimeric protein docking, Macromolecular docking, Protein–protein interactions, Protein–protein interface prediction

1 Introduction

Protein complexes are involved in many important processes in a living cell. In order to understand the mechanisms of these processes, it is necessary to solve the 3D structure of the protein complexes. Experimental techniques such as X-ray crystallography and nuclear magnetic resonance (NMR) have been used to solve the 3D structure of protein complexes, as shown in the large number of entries of complex structures in the Protein Data Bank (PDB) [1]. When protein complex structures have not been solved by experiments, it is possible to use computational tools to construct models of these complexes. A protein docking program takes two or more component protein structures as input and assembles

Daisuke Kihara (ed.), *Protein Structure Prediction*, Methods in Molecular Biology, vol. 1137,
DOI 10.1007/978-1-4939-0366-5_15, © Springer Science+Business Media New York 2014

them into 3D structure models of a protein complex. Input proteins can be either experimentally solved or computationally modeled structures using protein structure prediction programs such as the ones described in earlier chapters in this book.

Existing docking methods generate from a few hundred to thousands of candidate complexes, which are ranked by a score that indicates which models are more probable. Most of the existing docking methods deal with pairwise docking, where only two proteins are assembled [2–11]. A smaller number of methods perform docking of multiple protein structures, called multimeric or multiple-protein docking [12–17].

In this chapter we introduce three protein docking programs developed in our group. First, we show how to use LZerD (Local 3D Zernike Descriptor-based protein docking program) [11], our pairwise protein–protein docking program, to create protein complex structures from two individual proteins. It uses geometric hashing [18] for docking conformation search and a rotation invariant mathematical surface shape representation, the 3D Zernike Descriptors (3DZD) [19–22], as the main scoring term for evaluating docking poses. Next, we discuss PI-LZerD (Predicted Interface-guided LZerD) [23], which uses additional predicted protein interface information to guide conformation searches of pairwise protein–protein docking. PI-LZerD runs LZerD around the neighborhood of the provided predicted interface residues and further refines docking poses by running LZerD a second time. Finally, we present our multiple-protein docking program, Multi-LZerD [17, 24–26], which can assemble more than two proteins. In the first phase, pairwise docking predictions are generated for every possible pair of component proteins using LZerD. Then, multiple-protein complex structures will be generated by combining the pairwise docking predictions generated in the first phase. A genetic algorithm (GA) is used to explore the combinatorial space. After a configurable number of iterations, a final refinement step is applied to the structures.

The following sections provide instructions on how to use the LZerD protein–protein docking software. The readers are encouraged to refer to the original publications for more detailed descriptions of the algorithms and benchmark results of the programs.

2 Materials

The programs in the LZerD docking suite are available on the Kihara Lab website, http://www.kiharalab.org/proteindocking. The LZerD pairwise docking program is available as a compressed file (*lzerddistribution.tar.gz*) from the LZerD section of the webpage. Similarly, there is a section for PI-LZerD that has a link to the necessary files for PI-LZerD named *PI-LZerD.tar.gz*. Multi-LZerD is available as *multilzerddistribution.tar.gz* from the *Multi-LZerD* section. All packages are intended to run on Linux machines.

Once the files are downloaded to your computer, they need to be decompressed. If a graphical file explorer is used, right clicking on the file and choosing a decompression option should usually extract them. If you are using a command line terminal, you can decompress each file by running *tar –zxf lzerddistribution.tar.gz* (for the LZerD package, or specify the corresponding file name for PI-LZerD and Multi-LZerD). Once files are decompressed, a new folder will be created with several programs in it. The details of each package's contents, their roles in the procedure, and input data are described next.

2.1 Pairwise Docking Package, LZerD

LZerD needs two files containing protein structures as inputs. The two input files should follow the PDB format (the format is described at http://www.wwpdb.org/docs.html). LZerD only requires the ATOM fields in the file. Below, whichever protein is provided first will be called *receptor*, while the second one is called *ligand*.

The final output consists of many PDB files that represent different poses of the ligand only, since LZerD scans for poses of the ligand around the receptor placed at the original position. This means that the best docking pose needs to be generated by combining the receptor PDB file and ligand1.pdb, the file that contains the best ranking pose. In the same way, the second best prediction can be generated with the receptor file and ligand2.pdb, and so on. Optionally, the user can re-rank a subset of the docking poses using a physics-based scoring function, described later in the chapter.

A prediction can be executed with a series of programs written in C/C++ and a Linux Shell script that triggers the main process:

- runlzerd.sh: The main script that receives two input PDB files and carries out the complete process by invoking a series of programs.
- mark_sur: An auxiliary program that marks residues on the protein surface. It uses the uniCHARMM file that is also included in the package.
- GETPOINTS and LZD32: These two programs create surface information used by LZerD to compare protein surface shapes.
- LZerD1.0: The main program that performs the pairwise docking.
- PDBGEN: A post-processing program for generating the best ligand poses obtained by LZerD1.0.
- lzerd_rerank.sh: Script used to re-rank the docking poses using a physics-based scoring function.

2.2 Predicted Interface-Guided Protein Docking Package, PI-LZerD

To run PI-LZerD, the user needs the atomic structures of a receptor and a ligand in the PDB format as well as the list of predicted protein interface residues. Protein interface residues can be predicted using a protein interface prediction method, such as BindML [27] or meta_PPISP server [28], or they can be provided based on biological knowledge of the proteins. The output files represent predicted complex conformations in the PDB format. The files will be found in the PI_LZerD/10.Result directory.

2.3 Multiple Docking Package, Multi-LZerD

Similar to the previous program, Multi-LZerD receives protein structure files that follow the PDB format. Since Multi-LZerD is for multiple-protein docking, the number of files will be three or more, not just two. The input protein structure files should have a common file name and a suffix, ".pdb". Also, a prefix should be associated with the files to indicate the chain ID of the units. For example, three input files for a trimeric protein complex can be named as follows: A-mytrimer.pdb, B-mytrimer.pdb, and C-mytrimer.pdb. In this case the common file name is *mytrimer* and the prefixes are A, B, and C. In addition, it is required that the chain ID is provided in each ATOM field. We suggest that the same chain ID be used as the prefix.

The program outputs a series of PDB files. refined-00001.pdb is the best scoring model, refined-0002.pdb the second best, and so on. A set of decoy-<number>.pdb files are also created that represent the predicted structures before the refinement step (details described later). Multi-LZerD output files are complete models; i.e., each file contains all protein units in their predicted poses. This differs from LZerD where only the *ligand* part of the prediction is generated as output.

Since Multi-LZerD is a superset of LZerD, the programs mentioned in the previous LZerD section are also used here. The following list describes the additional scripts and programs that are incorporated for multiple docking:

- run.sh: The main script that executes the complete protocol. It is necessary to use proper suffix and prefixes for input file names before executing it.
- addhydrogens.pl: Given that we do not assume that the PDB files contain hydrogen atoms, this script will computationally add their atomic coordinates. This is required because of the scoring function used in Multi-LZerD.
- create_lzerd_decoys.pl and multilzerd_sort_decoys.pl: These two scripts trigger the pairwise docking performed by LZerD.
- multilzerd_pairwise_cluster.pl: The program performs clustering of pairwise docking predictions from LZerD to eliminate poses that are too similar.
- multilzerd: The main multiple protein docking program.

- multilzerd_create_pdb: Creates the decoy-<number>.pdb files that represent the final multiple docking poses.
- multilzerd_refine: An additional post-processing step that tries to improve the final multiple-protein docking poses.

3 Methods

This section describes steps to run the three docking programs: LZerD, PI-LZerD, and Multi-LZerD.

3.1 LZerD

The LZerD protocol flow chart is shown in Fig. 1. LZerD takes two PDB files, which will be docked with each other, as input.

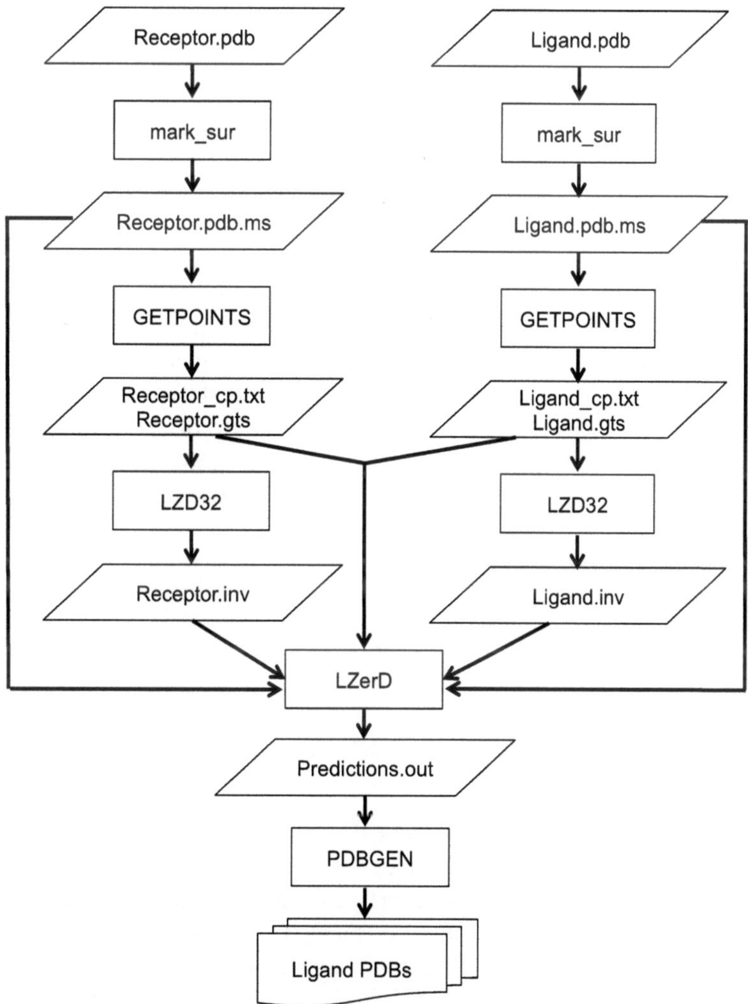

Fig. 1 LZerD flow diagram. Two input PDB files are first pre-processed using mark_sur, GETPOINTS and LZD32. Then, the intermediate files are fed into LZerD to create the pairwise docking predictions

They are called the receptor and ligand protein. The package contains a folder called *example*, which contains a sample receptor protein (1PPE_r_b.pdb) and a sample ligand (1PPE_l_b.pdb) protein file as well as all output files. We will use these two files to illustrate the procedure to perform protein docking with LZerD. We first show how to run the main script from a command line using default program settings. In the following sections, we explain each step in the process, intermediate files output by the programs, and possible parameter modifications users can control.

3.1.1 Main LZerD Process

In order to execute LZerD, users first need to locate the folder where the files were downloaded. For example, if the package was saved to a "Downloads" folder in the user's home directory, open a command terminal window and input the command

```
cd Downloads/lzerddistribution
```

Start the process by running the main script, *runlzerd.sh*, with two input PDB files as follows:

```
./runlzerd.sh example/1PPE_r_b.pdb example/1PPE_l_b.pdb
```

Here the sample receptor and ligand files provided in the example folder are used, but they can be any two PDB files. The process can run for several hours before reaching completion depending mainly on the size of the proteins. Some of the parameters used internally in the script will also influence how much computation time is required.

While the reader can change parameters in LZerD, simply running the shell script as shown above will complete the computation and produce candidate docking models. In the case of the example, the best model generated by LZerD will be composed of *1PPE_r_b.pdb* and *ligand1.pdb.* If the user requires a single PDB file, the two files can be concatenated by the *cat* command:

```
cat example/1PPE_r_b.pdb ligand1.pdb>model1. pdb
```

The new file *model1.pdb* can be visualized with a standard protein structure viewer such as PyMol or RasMol.

3.1.2 Detailed Steps

In the following sections, we provide more detailed information about each step in *runlzerd.sh* including parameters users can change.

Surface calculation. LZerD constructs the surface for two input protein structures in order to identify surface shape complementarity at the interface regions. This is the main term in the docking scoring function. First, it uses *mark_sur* to identify amino acid residues on the protein surface and adds annotations to the protein atoms in the ATOM fields of the PDB file. This will generate *.pdb.ms* files from the input *.pdb* files. In our example, *1PPE_r_b.pdb.ms* and *1PPE_l_b.pdb.ms* are the intermediate files created by this step:

```
./mark_sur 1PPE_r_b.pdb 1PPE_r_b.pdb.ms
./mark_sur 1PPE_l_b.pdb 1PPE_l_b.pdb.ms
```

Using marked surface residues, GETPOINTS creates two intermediate surface representations: an isosurface stored in *.gts* files and a point representation stored in *_cp.txt* files. The "-cut" parameter given to GETPOINTS is the key to control the execution time of the pairwise docking program as shown here:

```
./GETPOINTS -pdb 1PPE_r_b.pdb.ms -smooth 0.35 -cut 1e-04
./GETPOINTS -pdb 1PPE_l_b.pdb.ms -smooth 0.35 -cut 1e-04
```

This value is related to the precision of the surface representation used: lower values represent more fine-grained details and precision. The package downloaded contains a value of 1e-04, which we recommend for higher quality results with a long computation time. In contrast, using 1e-02 would yield a coarse-grained conformation sampling, which may be useful for obtaining preliminary results faster. A compromise between the time and the accuracy would be obtained by using a value of 1e-03.

Next, LZerD creates fingerprint representations of the shape around the surface points determined by *GETPOINTS*. These are stored in files with the *.inv* suffix. A fingerprint of a surface point is a vector of 36 numbers by default, which are coefficient values of a rotation invariant descriptor called the 3D Zernike descriptor [11, 22]. The *LZD32* program computes the fingerprints:

```
./LZD32 -g 1PPE_r_b.gts -c 1PPE_r_b_cp.txt -o 1PPE_r_b -dim 161 -rad 6.0 -ord 10
./LZD32 -g 1PPE_l_b.gts -c 1PPE_l_b_cp.txt -o 1PPE_l_b -dim 161 -rad 6.0 -ord 10
```

Please refer to our previous publications [21, 22] for more details about the 3D Zernike descriptors. The "-ord" parameter specifies the order of the 3D Zernike descriptors, which determines the number of elements in the vector. We found in our work [11] that 10 is an appropriate value, but the user could modify the order. For example, changing the order from 10 to 20 will increase the vector sizes from 36 to 121, which makes the fingerprint represent more details of the surface shape.

Pairwise docking. Using the files prepared in the previous step, LZerD1.0 will create protein–protein docking predictions:

```
./LZerD1.0 -rec 1PPE_r_b_cp.txt -lig 1PPE_l_b_cp.txt -prec
1PPE_r_b.pdb.ms -plig 1PPE_l_b.pdb.ms -zrec 1PPE_r_b_01.inv -zlig
1PPE_l_b_01.inv -rfmin 4.0 -rfmax 9.0 -rfpmax 15.0 -nvotes 8 -cor 0.7
-dist 2.0 -nrad 2.5 > 1PPE_r_b_1PPE_l_b. out
```

The output will be added to the file *1PPE_r_b_1PPE_l_b.out* progressively as the protein docking program analyzes different combinations of matching regions. The number of models generated can be obtained by counting the lines in the file, by running *wc* (word count):

```
wc 1PPE_r_b_1PPE_l_b.out
```

In a LZerD output file, each prediction is represented as a rotation and translation of the ligand protein from its original position. One line in the file is composed of 12 numbers, representing a transformation matrix, and an additional value that holds the score, for instance:

```
0.174 -0.868 0.465 -0.948 0.275 0.160 -0.267 0.413
-0.871 37.305 5.810 -2.871 397.116
```

In this example, the score is 397.116. The higher the score, the better the shape complementarity at the docking interface is, which indicates a better model.

In addition to this type of row it is possible that the file has a single header row like the following:

```
LIG: 0.926583 0.541623 2.32135 -1.96904 -8.93765 -2.37821
```

This is an optional random transformation that is applied to the ligand before beginning the docking process. This can be triggered by providing the –randomize flag when executing LZerD.

Generation of poses in PDB format. The number of models to output is one of the key parameters users can modify in *runlzerd.sh.* A program called PDBGEN is in charge of creating the different ligand docking poses in PDB format by receiving the receptor PDB, the ligand PDB, the output file, and the number of models to be created.

In the example case, PDBGEN is run as follows:

```
./PDBGEN 1PPE_r_b.pdb 1PPE_l_b.pdb 1PPE_r_b_1PPE_l_b.out 3
```

This generates three ligand files (ligand1.pdb, ligand2.pdb, and ligand3.pdb). The last number, 3, is the number of ligand models to generate, which can be increased to provide more models. PDBGEN can generate any number of models from one up to the number of lines in the ".out" file. As a reminder, only the ligand part of the complex is output in this step. To have a single PDB file with both receptor and ligand it is necessary to join both files.

Re-ranking LZerD predictions. The order in which rows appear in the *.out* file is based on the shape agreement between different docking poses. Optionally, users can re-rank the predictions based on a different type of score, a physics-based scoring function, described in one of our previous publications [17]. It considers van der Waals and electrostatic potentials, hydrogen and disulfide bonding, solvation, and additional knowledge-based contact terms. This is a more computationally intensive scoring method compared to the shape-based function used by LZerD. Thus, we recommend that this only be applied to a subset of the predictions in the *.out* file. To do so, the user needs to run the following script:

```
./lzerd_rerank.sh 1PPE_r_b.pdb 1PPE_l_b.pdb 1PPE_r_b_1PPE_l_b.out 100
```

This will generate a comma-separated file, called 1PPE_r_b_1PPE_l_b.out.reranked, with 100 text lines each representing a new score assigned to each of the first 100 predictions contained in the original *.out* file. This is a sample result that contains the first three lines of a re-ranked file:

```
model1.pdb,-9825.92,52,-69.0978,2112.1,
-19.1428,-1041.4,-207.964,1030.63,748.893,
-5.22232,-1403.18,-123.551,-14.0836
model21.pdb,-8636.75,41,-55.1373,1215.62,
-11.8204,-819.328,-189.976,805.575,665.787,
-2.24271,-1322.89,-123.589,-9.36367
model7.pdb,-6880.77,35,-43.6754,2579.88,
1.1024,-577.526,-111.942,554.141,252.346,
-4.48198,-1070.18,-123.285,-1.538
```

For example, the first line indicates that model1 (column 1) from the original *.out* file also has the best score using the physics-based function (column 2). However, we can see that the second best model, according to this new scoring function, comes from line 21 (model21) in the original *.out* file. The rest of the columns in the file represent the individual terms in the scoring function. Among them only the second column is used for re-ranking purposes. In this scoring function, a lower value represents a better score.

3.2 PI-LZerD

The execution of PI-LZerD takes a considerably longer time to finish compared to LZerD, because it runs pairwise docking multiple times. The PI-LZerD package consists of several programs, which are all combined in a shell script. First we present the shell script along with a reference to example files available online. Then, we discuss each step in the process and the estimated computational time required for the steps.

3.2.1 Main PI-LZerD Process

A single script, *PI_LZerD.sh*, runs the entire procedure from the beginning to the end. *PI_LZerD.sh* sets global variables based on user input and goes into each subdirectory to run individual programs sequentially. Each individual job receives parameters through global variables. For each job, *PI_LZerD.sh* checks the results and reports errors when necessary.

A detailed example is provided on the website using the protein complex (PDB ID: 1A2K) as an example (http://www.kiharalab.org/proteindocking/PI-LZerD/example). In this example, the input files are named 1A2K_R.pdb and 1A2K_L.pdb after their chain IDs, R and L, which are denoted in the PDB file. In addition, (predicted) docking interface residues should be provided. The files that contain the interface predictions in the example are called 1A2K_R.meta and 1A2K_L.meta. The user only needs to execute the following command in a terminal to run the complete PI-LZerD protocol:

```
./PI_LZerD.sh 1A2K R L
```

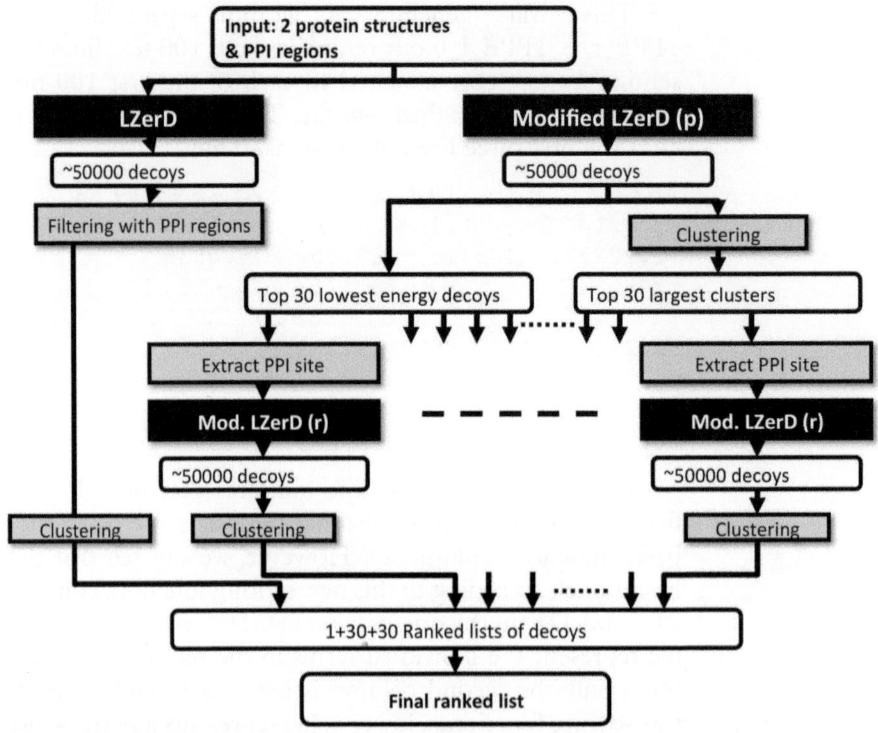

Fig. 2 Overview of the PI-LZerD algorithm. On the left branch the original LZerD program is run. LZerD (p) uses the permissive search space, and LZerD (r) uses restrictive search space employed in the geometric hashing stage. Refer also to the original PI-LZerD paper [23]

3.2.2 Detailed Steps

The flow chart of the PI-LZerD algorithm (Fig. 2) is provided in the original PI-LZerD paper [23]. PI-LZerD is computationally expensive, because it runs LZerD pairwise docking 61 times in an iterative fashion. Thus, the total running time can be over a week, if executed sequentially. The 60 independent runs of LZerD in the second iteration can be distributed across multiple CPUs in a single machine or in a computing cluster and run in parallel.

The PI-LZerD package is organized in a folder structure that reflects computational steps (Fig. 3). There are 16 major steps in total as shown in Fig. 4, which include 23 sub-steps listed in Table 1. For each folder, a script named *job.sh* is used to run the corresponding programs. The *job.sh* scripts communicate with the main script, *PI_LZerD.sh*, using the global environment with the main script running the scripts in each directory following the numeric order. It also checks the return values from each script and detects if any errors occurred. As shown in Fig. 4, the whole procedure can be divided into five stages.

Stage 1. LZerD run, without interface residue information: LZerD is run without using protein interface information. This stage corresponds to the leftmost branch in Fig. 2. The LZerD program used in PI-LZerD has been modified from the original LZerD program. The modified LZerD outputs matched critical points from two proteins in each predicted conformation. Files are compressed and saved to a file called <PDB_ID> .out.gz. This information is later used for both the permissive LZerD and restrictive LZerD

```
LZerD/                          PI_LZerD
|-- 01.BASE                     |-- 01.Pred_PPI
|    |-- CP                     |    |-- 01.Pred_PPI_pdb
|    |-- INV                    |    |-- 02.PPI_CP
|-- 02.MPI                      |    |-- 03.PPI_RES
|    |-- 01.LZerD_MPI           |    `-- 04.Res_Dist
|    |-- 02.LZerD_MAT           |-- 02.Sim_LZerD
|    |-- 03.SRB_MPI             |-- 03.LZerD_CIRMSD
|    `-- 04.IRMSD_MPI           |-- 04.CIRMSD_CLUST
|-- 03.Votes                    |-- 05.CLUST_T60
|-- 04.ORD_SRB                  |-- 06.SRF
|-- 05.CI_RMSD                  |-- 07.T60_Iter
`-- 06.LZerD_CIRMSD             |-- 08.K60_CIRMSD
                                |-- 09.T61
                                `-- 10.Result
```

Fig. 3 PI-LZerD directory structure. There are two directories in PI-LZerD: LZerD and PI_LZerD. Under each directory, there is a script named job.sh, which is sequentially called by the main script PI_LZerD.sh

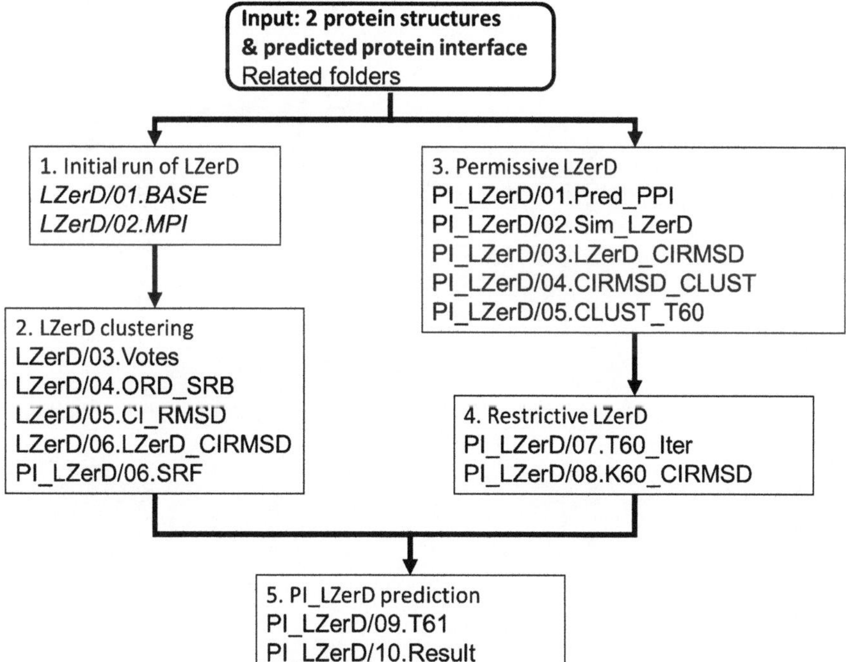

Fig. 4 Computing steps in PI-LZerD. Five steps of PI-LZerD and programs included in the steps

Table 1
Detailed steps of PI-LZerD. All 23 steps in PI-LZerD are summarized using 1A2K as an example of input. A brief description and input and output files are provided

Stage	Step	Description	Input	Input example	Output	Output example
1	`LZerD/ 01.BASE`	Add hydrogen atoms, find and mark the surface residues using mark_sur	Receptor and ligand protein structure files	`1A2K_R.pdb, 1A2K_L.pdb`	Receptor and ligand protein structures with flagged surface residue information	`1A2K_R.pdb, 1A2K_L.pdb`
1	`LZerD/ 01.BASE/CP`	Generate critical point information using GETPOINTS	Receptor and ligand proteins with surface residue information	`1A2K_R.pdb, 1A2K_L.pdb`	Critical pint (.cp) files	`1A2K_R.gts, 1A2K_R.cp, 1A2K_L.gts, 1A2K_L.cp`
1	`LZerD/ 01.BASE/INV`	Compute 3D Zernike descriptors for each critical point	gts and critical point (.cp) files	`1A2K_R.gts, 1A2K_R.cp, 1A2K_L.gts, 1A2K_L.cp`	Zernike descriptors in inv format	`1A2K_R.inv, 1A2K_L.inv`
1	`LZerD/ 02.MPI/ 01.LZerD_MPI`	Initial run of protein docking prediction using LZerD	Receptor and ligand structure, critical point information, Zernike descriptors	`1A2K_R.pdb, 1A2K_L.pdb, 1A2K_R.gts, 1A2K_R.cp, 1A2K_L.gts, 1A2K_L.cp`	Initial run of protein docking predictions	`1A2K.out.gz`
1	`LZerD/ 02.MPI/ 02.LZerD_MAT`	Extract rotation matrices information for physics-based scoring and IRMSD computation	Extract rotation metrics information for physics-based scoring and IRMSD computation	`1A2K.out.gz`	Extracted rotation/translation matrices information	`1A2K.mat`
1	`LZerD/ 02.MPI/ 03.SRB_MPI`	Compute physics-based score for each prediction	Rotation/translation matrices, and the receptor–ligand protein structure	`1A2K.mat`	Physics-based scores	`1A2K.srb`

1	LZerD/02.MPI/04.IRMSD_MPI	Compute physics-based interface RMSD (IRMSD) for each prediction	Rotation/translation matrices, and the receptor–ligand protein structure	1A2K.mat	Interface RMSDs	1A2K.irmsd
1	LZerD/03.Votes	Extract the critical point dependencies	Initial protein docking prediction result	1A2K.out.gz	The critical point dependency information	1A2K.Votes
2	LZerD/04.ORD_SRB	Sort predictions by physics-based scores	Initial protein docking prediction result	1A2K.srb, 1A2K.irmsd	Sorted predictions	1A2K.LZerD
2	LZerD/05.CI_RMSD	Compute pairwise Common Interface RMSD (CI_RMSD) for top 1,000 predictions	Initial protein–protein docking predictions	1A2K.LZerD	Common Interface RMSD (CI_RMSD) for top 1,000 predictions	1A2K.cirmsd
2	LZerD/06.LZerD_CIRMSD	Cluster predictions based on CI_RMSD	Top 1,000 predictions	1A2K.LZerD, 1A2K.cirmsd	Clustered top 1,000 predictions	1A2K.t1k
2	PI_LZerD/06.SRF	Use naïve-filtering method on predicted interface residues	Top 1,000 predictions from first iteration LZerD program	1A2K.LZerD, res.txt	Top 1,000 predictions sorted by the percentage of agreement with the predicted interface residues	1A2K.rcf
3	PI_LZerD/01.Pred_PPI/01.Pred_PPI_pdb	Protein interface prediction results from meta-PPISP server	Receptors and ligand structure files	1A2K_R.meta, 1A2K_L.meta	Predicted protein interface	1A2K_R.rec.pdb, 1A2K_L.lig.pdb
3	PI_LZerD/01.Pred_PPI/02.PPI_CP	Compute the critical points belonging to predicted interface residues	Receptor/ligand structure and predicted protein interfaces	1A2K_R.cp, 1A2K_L.cp, 1A2K_R.rec.pdb, 1A2K_L.lig.pdb	Critical points on predicted interface residues	1A2K_R.rec.pdb, 1A2K_L.lig.pdb

(continued)

Table 1 (continued)

Stage	Step	Description	Input	Input example	Output	Output example
3	PI_LZerD/01. Pred_PPI/03. PPI_RES	Compute the predicted interface residues	Predicted protein interface	1A2K_R.cp, 1A2K_L.cp, 1A2K_R.rec.pdb, 1A2K_L.lig.pdb	List of predicted interface residues	res.txt
3	PI_LZerD/02. Sim_LZerD	Permissive LZerD using predicted interface residues	Receptor and ligand structure and list of predicted interface residues	1A2K.Votes, res.txt	First iteration LZerD prediction	1A2K.sim
3	PI_LZerD/03. LZerD_CIRMSD	Compute pairwise CI_RMSD distances on top 1,000 predictions	Top 1,000 predictions from initial run of LZerD prediction	1A2K.sim	CI_RMSD distances of top 1,000 predictions	1A2K.cirmsd
3	PI_LZerD/04. CIRMSD_CLUST	Cluster top 1,000 predictions based on CI_RMSD distances	Top 1,000 predictions from initial run of LZerD program	1A2K.cirmsd	Clustering from top 1,000 predictions	1A2K.t1k
3	PI_LZerD/05. CLUST_T60	Select top 60 clustered predictions	Clustered top 1,000 predictions	1A2K.t1k	Selected top 60 clustered predictions	1A2K.40A.t60
4	PI_LZerD/07. T60_Iter	Restrictive LZerD using 60 clustered prediction	Selected top 60 clustered predictions	1A2K.40A.t60	Restrictive LZerD prediction based on the 60 clustered predictions	1A2K.iter.gz, 1A2K.k60
4	PI_LZerD/08. K60_CIRMSD	Pairwise CI_RMSD distances on top 1,000 predictions for 60 prediction lists	Top 60 clustered predictions	1A2K.k60	60× Pairwise CI_RMSD distances	1A2K.cirmsd
5	PI_LZerD/09. T61	Merge lists of 61 predictions	Selected top 60 clustered predictions	1A2K.cirmsd, 1A2K.LZerD	Re-ranked predictions	1A2K.i61
5	PI_LZerD/10. Result	Generate complex conformation for each prediction	Merged prediction list	1A2K.i61	Complex conformation for each prediction	1A2K.1.pdb

process (the right branch in Fig. 2). A score is computed each predicted conformation with a physics-based scoring function, the same function used in the LZerD pairwise docking. Folders for Step 1 and their roles are as follows:

Related folders:

```
LZerD/01.BASE Add hydrogen atoms, compute critical points and 3DZD for input PDB files
LZerD/02.MPI  Initial run of LZerD
LZerD/03.Votes Compute matching critical points for each prediction
LZerD/04.ORD_SRB Computing physics-based score
```

Stage 2. Clustering docking models: The top 1,000 scoring docking models computed in Stage 1 are clustered. The similarity of docking models is measured with the Common Interface RMSD (CI_RMSD), which is the RMSD computed only for residues that are common in the models to be compared. Then, the representatives of the clusters are sorted by considering the agreement of the docking interface residues to the provided predicted protein interface information (naïve-filtering method). This corresponds to the clustering of the leftmost branch in Fig. 2.

Related folders:

```
LZerD/05.CI_RMSD          Compute CI_RMSD
LZerD/06.LZerD_CIRMSD     Clustering by CI_RMSD
PI_LZerD/06.SRF       Naïve-filtering method
```

Stage 3. LZerD run with provided docking interface residue information (permissive search). Next, the process moves to the right branch in Fig. 2. LZerD is run to explore docking conformations in the vicinity of the provided docking interface residue information. More concretely, the interaction interface of docking poses needs to have some overlap with the provided interface residues. The permissive search means that the conformation search space is set to be larger compared to the next iteration of LZerD runs (Stage 4). 50,000 docking models are generated and clustered in terms of CI_RMSD.

Related folders:

```
PI_LZerD/01.Pred_PPI        Predicted protein interface information
PI_LZerD/02.Sim_LZerD       Running Permissive LZerD
PI_LZerD/03.LZerD_CIRMSD  CI_RMSD of permissive LZerD results
PI_LZerD/04.CIRMSD_CLUST  Clustering of permissive LZerD results
```

PI_LZerD/05.CLUST_T60 Selecting Top 60 cluster selection

Stage 4. Second LZerD run with interface residue information (restrictive search): Among the clustering results of the previous LZerD (Stage 3), the 30 largest clusters are selected and the docking model that is closest to the cluster centroid in each cluster is kept. In addition, the 30 lowest energy docking models are also selected (the energy values of all docking models have already been computed and stored in Stage 1). Thus, a total of 60 docking models are kept from Stage 3. For each of the 60 docking models, docking interface residues are extracted and considered as updated predictions of interface residues. Then, LZerD is run using the updated interface residue information similar to Stage 3, but this time the conformational search is restricted to a smaller area close to the provided interface residues. Docking models are clustered based on CI_RMSD, and the one model that is closest to the cluster centroid is kept.

Related folders:

PI_LZerD/07.T60_Iter Restrictive LZerD

PI_LZerD/08.K60_CIRMSD Clustering with CI_RMSD for restrictive LZerD results

Stage 5. Generating PI-LZerD prediction: At this stage there are 61 ranked lists of docking models, each from an independent LZerD run. The last step of PI-LZerD is to compute the final ranked list of predictions out of the 61 lists. From each of the 61 lists, the top-ranked predictions are first ranked among themselves by their scores. Then, the second-ranked predictions from each file are ranked in the same way. This is repeated for predictions at the same subsequent ranks in the files. Thus, the final output is a list of predictions in the 61 lists, which are first sorted by their ranks and then sorted by the scores. Finally, the conformations of the top predictions are then generated.

Related folders:

PI_LZerD/09.T61 61 prediction lists

PI_LZerD/10.Result Merged 61 prediction lists

Table 2 lists the expected running time for each step. Generally a "fast" process can be finished in minutes, a "medium" process should be finished in hours, and a "slow" process can take days to finish.

3.3 Multi-LZerD

The Multi-LZerD package contains a script, *run.sh*, that runs the complete process of multiple-protein docking. The overall steps are illustrated in Fig. 5. To execute *run.sh*, the names of the input files need to be specified as explained in the next section. The later sections will go into more details about the intermediate steps performed by *run.sh*.

Table 2
Running time for steps in PI-LZerD. Fast indicates minutes, medium indicates hours, and slow processes can take days to finish

Stage	Folder	Time	References
1	LZerD/01.BASE	Medium	Add hydrogen atoms, create critical points, compute Zernike descriptors
1	LZerD/02.MPI/01.LZerD_MPI	Slow	Original LZerD program with modified output
1	LZerD/02.MPI/02.LZerD_MAT	Medium	Extract prediction matrix
1	LZerD/02.MPI/03.SRB_MPI	Slow	Compute the physics-based scores
1	LZerD/02.MPI/04.IRMSD_MPI	Slow	Compute the interface RMSD
1	LZerD/03.Votes	Medium	Extract the critical point dependence information
2	LZerD/04.ORD_SRB	Medium	Sort predictions in physics-based scores
2	LZerD/05.CI_RMSD	Medium	Compute common interface RMSDs (CI-RMSD)
2	LZerD/06.LZerD_CIRMSD	Medium	Generate list of clustered original LZerD predictions
2	PI_LZerD/06.SRF	Medium	Apply simple residue filtering method
3	PI_LZerD/01.Pred_PPI/	Fast	Extract protein interface information
3	PI_LZerD/02.Sim_LZerD	Slow	LZerD(p)
3	PI_LZerD/03.LZerD_CIRMSD	Medium	Compute CI-RMSDs for LZerD(p)
3	PI_LZerD/04.CIRMSD_CLUST	Medium	Cluster based on CI-RMSD
3	PI_LZerD/05.CLUST_T60	Medium	Generate top 30 + 30 predictions
4	PI_LZerD/07.T60_Iter	Slow	Run LZerD(r) for 60 predictions
4	PI_LZerD/08.K60_CIRMSD	Medium	Compute CI-RMSD for 60 predictions
5	PI_LZerD/09.T61	Fast	Generate list from 60 + 1 lists of predictions
5	PI_LZerD/10.Result	Fast	Generate complex conformations

3.3.1 Main Shell Script for Multi-LZerD

Multi-LZerD requires that input file names follow a specific naming convention. As an example, we provide three protein files in the package: A-sample.pdb, B-sample.pdb, and C-sample.pdb. Every file name must start with a prefix which indicates the chain ID of the protein. We recommend using the same prefix as the chain

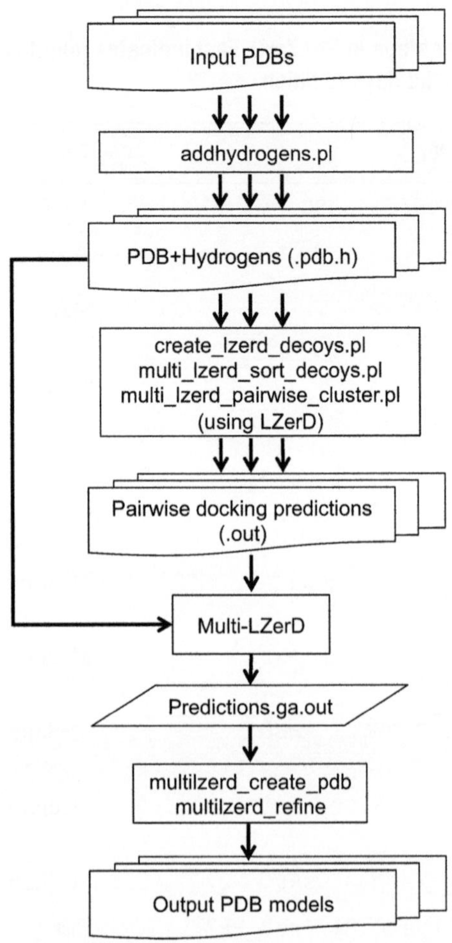

Fig. 5 Multi-LZerD flow diagram. Several PDB files, each representing a protein unit, are pre-processed to add hydrogen information and to generate pairwise docking predictions using LZerD. Multi-LZerD then combines the pairwise predictions and generates PDB files that contain the multimeric protein models

identifier in the PDB file of the protein. The prefix should be followed by a "-" and a base name (e.g., PDB file name, e.g., 1abc or *sample* in the case of the sample files) and a suffix, *.pdb*.

Once the input PDB file names are modified following the rule above, the file names should be specified in *run.sh* as follows:

```
basename = 'sample'
units = 'A,B,C'
```

Notice that the "-" and the suffix *.pdb* are not included in either the unit prefixes or as base name. Also, the unit prefixes must be comma separated. If the names of the files were *D-complex.pdb*, *E-complex.pdb*, and *F-complex.pdb*, they would be specified as follows:

```
basename = 'complex'
units = 'D,E,F'
```

Once the input files are specified in *run.sh*, users can run it by opening a terminal window and navigating to the folder where *run.sh* is located by using the *cd* command, and then, simply run it by typing:

```
./run.sh
```

When the computation is finished, a series of files with names starting with "refined" are created, which contain predicted protein complex models. The associated number indicates the rank of the models, with lower numbers representing better predictions.

3.3.2 Detailed Steps in Multi-LZerD

Here we explain the steps in *run.sh*. Parameters in some steps can be modified.

1. *Addition of hydrogen atoms.* The scoring function used in Multi-LZerD requires the presence of hydrogen atoms in protein structures. A script named *addhydrogen.pl* takes PDB files and generates a new set of files with a ".pdb.h" suffix:

```
./addhydrogens.pl sample A,B,C
```

   The newly created files contain all the original atoms and, additionally, hydrogen atom locations calculated using the HBPLUS program [29]. If users prefer to add hydrogen atoms in an alternative way, this step should be removed from *run.sh*. Please note that the files with hydrogen atoms should have a file name with *.pdb.h* at the end.

2. *Pairwise docking predictions.* Multi-LZerD uses LZerD to create pairwise poses between all possible combinations of pairs of component proteins. This means that LZerD will be executed several times. The following section in *run.sh* manages the necessary calls to LZerD to create pairwise docking predictions:

```
./create_lzerd_decoys.pl sample A,B,C execute
```

   create_lzerd_decoys.pl will terminate once all pairwise predictions are finished. Then, the top predictions for each case are kept using the following auxiliary script:

```
./multilzerd_sort_decoys.pl ./A,B,C 54000
```

   The number of pairwise poses kept in this example is 54,000, although users can change this value. A lower number will reduce the search space of complex structures and the execution time but may eliminate correct predictions. A higher number will increase the chance of finding correct and near-correct poses but at the price of increasing the execution time. These steps will produce result files that contain pairwise docking predictions for each pair. The file names will be, for the example case, *A-B.out*, *A-C.out*, and *B-C.out*. In general, there will be a "*.out*" file for every pair of proteins identified by their prefixes.

3. *Removal of similar pairwise predictions.* By default, the protocol generates 54,000 pairwise poses per protein pair, but a number of them may be similar to each other. To minimize redundancy we apply a clustering program to group similar poses and keep one representative pose from every group. Predictions are grouped together if their root mean square deviation (RMSD) between C-α atoms is less than 10 Å, in the default case. Changing the following line in *run.sh* modifies this threshold:

```
cluster_threshold = 10
```

More pairwise poses will be kept with a smaller value. Conversely, a higher value will reduce pairwise docking conformations and the execution time more aggressively but with a risk of pruning out correct or near-correct models.

4. *Multiple-protein docking.* The main multiple-protein docking is performed by the following line in *run.sh*:

```
./multilzerd --pdbid sample --chains A,B,C
-o sample --generations 200
  --population 200 --clashes 300 --cluster 10
--weights all
```

Users may want to change four parameters provided to the program, depending on the protein complexes they want to assemble.

5. *Number of generations.* The generation parameter, which is set to `--generations 200` as default, represents the number of iterations used to explore different pairwise prediction combinations during the GA performed by Multi-LZerD. More conformations will be explored with a higher number of iterations. The amount of iterations required varies for different cases. In our papers, we have explored the effect that the number of generations has in the final prediction accuracy [17, 24]. For cases up to six chains we observed that high-quality models were obtained within 2,000–3,000 iterations. Notice that the script sets the number of iterations to 200. We recommend this number to be increased if users can bear a longer computation time.

6. *Population size.* The population size, set as `--population 200`, is the number of conformations generated and compared in each generation of GA. A larger population size can explore more conformations. Obviously, the execution time will increase as the population increases. We have analyzed how much the population size impacts the quality of the models in our previous publication [24]. For most cases we recommend a population size between 200 and 400.

7. *Number of atom clashes allowed.* The atom clash parameter (--clashes), which is set to 300 as default, determines the number of atoms that we allow to be closer than 3 Å to another atom in a protein complex model. In the default setting, if 300 or more atoms in a model have clashes, the model is removed. When a target protein complex has a larger number of subunits, more clashes may be tolerated. The default setting of 300 clashes is appropriate for a smaller number of chains (e.g., 3 or 4). We have found that it needs to be increased to around 800–1,000 for a six-chain complex.

8. *Multiple-docking clustering threshold.* We previously mentioned a clustering threshold for pairwise docking. Clustering is also performed against a population of protein complexes at the end of every GA generation to remove redundant models. The RMSD between C-α atoms of models is also used in this case, and it is set by the --cluster flag. The default value is set to 10 Å. By default, the cutoffs of both pairwise and multiple docking are set to 10 Å. We would recommend users to keep it at 10 Å, but they do not need to be the same value.

9. *Generation of poses in the PDB format.* After Multi-LZerD finishes, a file with a *.ga.out* suffix will be generated. It will contain the same number of models as the number provided for the --population parameter. To generate PDB files from the output file, *run.sh* executes the following command:

```
./multilzerd_create_pdb sample.ga.out ./1 200 decoy
```

This will create 200 PDB files that start with the "decoy-" prefix, followed by a number that shows the model rank. This program receives as input five parameters:

- A *.ga.out* output file.
- The directory where the pairwise output files and the original PDB files are located. In the example provided, this means that *A-sample.pdb*, *B-sample.pdb*, *C-sample.pdb*, *A-B.out*, *A-C.out*, and *B-C.out* are in the current directory "./".
- Two index numbers between 1 and the population size used that determine the rank of the PDB models to output. In the example, since the population size used was 200, 1 and 200 imply that all the models will be generated. An alternative range such as 25 to 50 means that the first 24 models will not be generated and models between 25 and 50 (inclusive) will.
- A prefix that will be used in all of the output file names (*decoy* in the example).

10. *Refinement.* It is possible to apply a refinement program in the package to the generated models. The refinement program will perform small translations and rotations between subunits in a complex model to find a complex structure of lower energy.

Resulting models are not expected to deviate significantly from the starting conformation; rather, the program is designed for minor adjustments. This is achieved by the last command in *run.sh*:

```
./multilzerd_refine --input sample.ga.out
--trials 200 --prefix refined
```

The refinement tries a configurable number of randomized modifications, specified by the *--trials* flag. While we provide 200 as a default value, we have used 2,000 in some of our previous works. One refined model will be created for each model in the *.ga.out* file. Refined PDB files have the following naming convention: refined- < ##### > .pdb (with a five-digit number that corresponds to the prediction number in the *.ga.out* file).

4 Case Studies

Here we show examples of predicted protein complexes by LZerD, PI-LZerD, and Multi-LZerD. Figure 6a is a pairwise docking prediction by LZerD for a protein complex of a bovine beta-trypsin and an inhibitor CMTI-I (PDB ID: 1PPE). This is one of the entries in a protein–protein docking benchmark set [30]. The results for a sample LZerD [11] run are visualized with PyMOL [31] in part A of this figure. Two predicted positions for the ligand, CMTI-I, are shown. The structure in blue corresponds to *ligand1.pdb* (iRMSD 14.3 Å), while the best prediction, ranked 505 in this sample run, is shown in orange (iRMSD 1.12 Å).

The second example (Fig. 6b) is a prediction by PI-LZerD for a complex of human CDK2 kinase with cell cycle regulatory protein CksHs1. The correct docking pose is shown in blue, and the prediction by PI-LZerD is shown in green (iRMSD: 1.03 Å). The interface residue prediction used has a sensitivity (fraction of correctly predicted interface residues) of 0.33 for the receptor and 0.44 for the ligand. Interface residue information can also be used as post-screening for docking models, where models that have the same set of interface residues as the provided interface residues are selected among all of the produced docking models. The post-screening (called naïve-filtering method) did not work well for this case because the interface residue information is not very accurate (predicted pose shown in red).

Finally, we show a prediction example of Multi-LZerD (Fig. 6c). We have used a trimeric complex of two transcription factors and a retinoblastoma-associated protein (PDB ID: 2AZE). This trimer has a triangular topology, which means that all three units interact with the other two. A near-native prediction (a global C-α RMSD of 0.99 Å) was found as the lowest energy model as

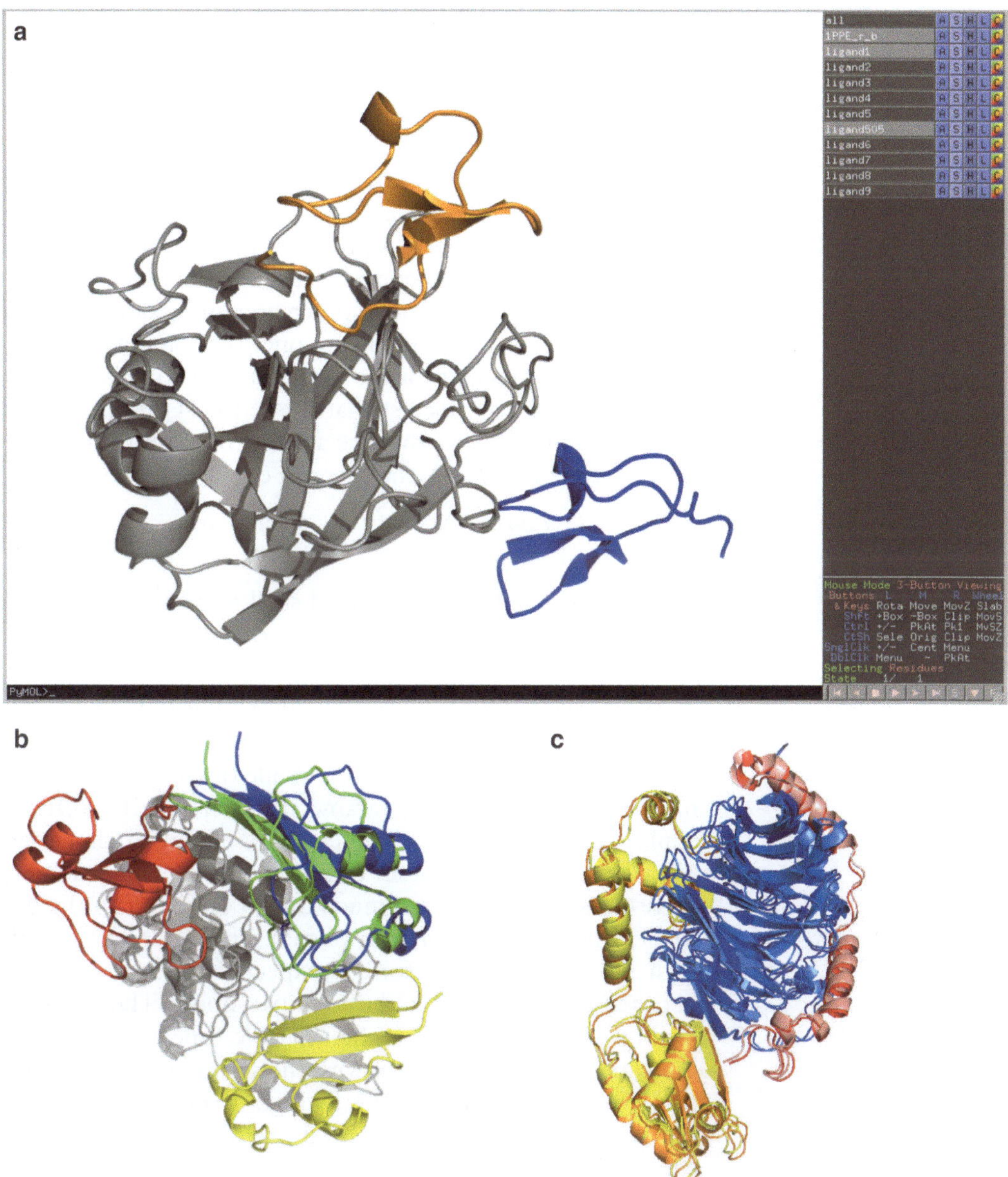

Fig. 6 Visualization of LZerD case studies. (**a**) Snapshot of visualization of predicted docking models for a protein complex (PDB ID: 1PPE) using PyMOL. Prediction 1 (*blue*) and 99 (*orange*) are visualized, while the other models are in the environment but not shown. (**b**) Predictions generated by PI-LZerD for 1BUH. The native structure is shown in *blue*, PI-LZerD's prediction in *green*, the standard LZerD in *yellow*, and the *red one* shows a prediction using the naïve-filtering method, which simply selects models by examining the consistency of interface residues in models with predicted interface residues. This figure is modified from a figure originally published in the PI-LZerD paper [23]. (**c**) A Multi-LZerD prediction for 1A0R. *Blue*, *red*, and *yellow* show the native conformation, while *light blue*, *salmon*, and *orange* show the predicted unit poses

shown in the original Multi-LZerD publication [17]. The native units are shown in blue, red, and yellow colors, while the predicted units are colored using light blue, salmon, and orange.

5 Notes

In this section we provide a few hints that we have found useful when executing our docking protocols.

1. *Running LZerD and Multi-LZerD with nohup*: We encourage users to execute either LZerD or Multi-LZerD using nohup, a Linux command that allows processes to keep running even if users log out from a computer. This is useful especially when users remotely log into a computer because it allows users to start a docking process and log out once the command is issued without having to wait for the whole process to finish.

   In addition, we suggest that the terminal output be redirected to a file, which can serve as a log of the execution. For example, to run LZerD, the user can execute this command:

```
nohup ./runlzerd.sh example/1PPE_r_b.pdb example/1PPE_l_b.pdb >& log.txt &
```

   The ">&" tells the command shell to redirect all output, which are normally shown on the terminal screen to *log.txt*. Notice that not all command shells use the same nomenclature to redirect the output. The final ampersand symbol tells the terminal to let the program run in the background.

2. *Optimizing GETPOINTS cutoff*: While discussing LZerD we mentioned that the *-cut* parameter given to GETPOINTS could vary the runtime significantly. We suggest that users, when testing a new protein complex, first set this parameter to 1e-02 and observe how much time it takes to run the protocol. Then run the program again with 1e-03 and finally with 1e-04. The first setting is expected to have a runtime between minutes to around one to two hours. The latter two settings take more time. If the available time allows it, users can run the procedure again with the second or the third setting. If the new setting is taking longer than users can afford, the new run can be aborted, and the results from the previous run can be used.

Acknowledgments

The authors thank Kristen Johnson for proofreading the manuscript. This work has been supported by grants from the National Institutes of Health (R01GM075004 and R01GM097528), National Science Foundation (EF0850009, DBI1262189, IOS1127027, IIS1319551), and National Research Foundation of Korea Grant funded by the Korean Government (NRF-2011-220-C00004). J.E.R. would like to thank the Fulbright Science and Technology program for supporting his first years of graduate studies.

References

1. Rose PW, Bi C, Bluhm WF et al (2013) The RCSB Protein Data Bank: new resources for research and education. Nucleic Acids Res 41:D475–D482. doi:10.1093/nar/gks1200
2. Ben-Zeev E, Eisenstein M (2003) Weighted geometric docking: incorporating external information in the rotation-translation scan. Proteins 52:24–27. doi:10.1002/prot.10391
3. Chen R, Li L, Weng Z (2003) ZDOCK: an initial-stage protein-docking algorithm. Proteins 52:80–87. doi:10.1002/prot.10389
4. Dominguez C, Boelens R, Bonvin AMJJ (2003) HADDOCK: a protein-protein docking approach based on biochemical or biophysical information. J Am Chem Soc 125:1731–1737. doi:10.1021/ja026939x
5. Gray JJ, Moughon S, Wang C et al (2003) Protein-protein docking with simultaneous optimization of rigid-body displacement and side-chain conformations. J Mol Biol 331:281–299
6. Moreira IS, Fernandes PA, Ramos MJ (2010) Protein-protein docking dealing with the unknown. J Comput Chem 31:317–342. doi:10.1002/jcc.21276
7. Pierce B, Weng Z (2007) ZRANK: reranking protein docking predictions with an optimized energy function. Proteins 67:1078–1086. doi:10.1002/prot.21373
8. Ritchie DW (2008) Recent progress and future directions in protein-protein docking. Curr Protein Pept Sci 9:1–15
9. Schneidman-Duhovny D, Inbar Y, Nussinov R, Wolfson HJ (2005) PatchDock and SymmDock: servers for rigid and symmetric docking. Nucleic Acids Res 33:W363–W367. doi:10.1093/nar/gki481
10. Tovchigrechko A, Vakser IA (2006) GRAMM-X public web server for protein-protein docking. Nucleic Acids Res 34:W310–W314. doi:10.1093/nar/gkl206
11. Venkatraman V, Yang YD, Sael L, Kihara D (2009) Protein-protein docking using region-based 3D Zernike descriptors. BMC Bioinforma 10:407. doi:10.1186/1471-2105-10-407
12. André I, Bradley P, Wang C, Baker D (2007) Prediction of the structure of symmetrical protein assemblies. Proc Natl Acad Sci USA 104:17656–17661. doi:10.1073/pnas.0702626104
13. Inbar Y, Benyamini H, Nussinov R, Wolfson HJ (2005) Prediction of multimolecular assemblies by multiple docking. J Mol Biol 349:435–447. doi:10.1016/j.jmb.2005.03.039
14. Berchanski A, Eisenstein M (2003) Construction of molecular assemblies via docking: modeling of tetramers with D2 symmetry. Proteins 53:817–829. doi:10.1002/prot.10480
15. Comeau SR, Camacho CJ (2005) Predicting oligomeric assemblies: N-mers a primer. J Struct Biol 150:233–244. doi:10.1016/j.jsb.2005.03.006
16. Karaca E, Melquiond ASJ, De Vries SJ et al (2010) Building macromolecular assemblies by information-driven docking: introducing the HADDOCK multi-body docking server. Mol Cell Proteomics 9:1784–1794. doi:10.1074/mcp.M000051-MCP201
17. Esquivel-Rodríguez J, Yang YD, Kihara D (2012) Multi-LZerD: multiple protein docking for asymmetric complexes. Proteins 7:1818–1833. doi:10.1002/prot.24079
18. Wolfson HJ, Rigoutsos I (1997) Geometric hashing: an overview. IEEE Comput Sci Eng 4:10–21. doi:10.1109/99.641604
19. Canterakis N (1999) 3D Zernike moments and Zernike affine invariants for 3d image analysis and recognition. 11th scandinavian conference on image analysis
20. Novotni M, Klein R (2003) 3D zernike descriptors for content based shape retrieval. Proceedings of the eighth ACM symposium on solid modeling and applications—SM'03. ACM Press, New York, NY, USA, p 216
21. Kihara D, Sael L, Chikhi R, Esquivel-Rodríguez J (2011) Molecular surface representation using 3D Zernike descriptors for protein shape comparison and docking. Curr Protein Pept Sci 12:520–530, doi: http://dx.doi.org/10.2174/138920311796957612
22. Sael L, Kihara D (2009) Protein surface representation and comparison: new approaches in structural proteomics. In: Chen JY, Lonardi S (eds) Biological data mining. Chapman & Hall/CRC, Boca Raton, FL, pp 89–109
23. Li B, Kihara D (2012) Protein docking prediction using predicted protein-protein interface. BMC Bioinforma 13:7. doi:10.1186/1471-2105-13-7
24. Esquivel-Rodríguez J, Kihara D (2012) Effect of conformation sampling strategies in genetic algorithm for multiple protein docking. BMC Proc 6 Suppl 7:S4. doi: 10.1186/1753-6561-6-S7-S4
25. Esquivel-Rodríguez J, Kihara D (2012) Fitting multimeric protein complexes into electron microscopy maps using 3D Zernike descrip-

tors. J Phys Chem B 23:6854–6861. doi:10.1021/jp212612t

26. Esquivel-Rodríguez J, Kihara D (2012) Evaluation of multiple protein docking structures using correctly predicted pairwise subunits. BMC Bioinforma 13:S6. doi:10.1186/1471-2105-13-S2-S6
27. La D, Kihara D (2012) A novel method for protein-protein interaction site prediction using phylogenetic substitution models. Proteins 80:126–141. doi:10.1002/prot.23169
28. Qin S, Zhou H-X (2007) meta-PPISP: a meta web server for protein-protein interaction site prediction. Bioinformatics 23:3386–3387. doi:10.1093/bioinformatics/btm434
29. McDonald IK, Thornton JM (1994) Satisfying hydrogen bonding potential in proteins. J Mol Biol 238:777–793. doi:10.1006/jmbi.1994.1334
30. Hwang H, Vreven T, Janin J, Weng Z (2010) Protein-protein docking benchmark version 4.0. Proteins 78:3111–3114. doi:10.1002/prot.22830
31. Schrödinger L (2010) The PyMOL Molecular Graphics System, Version 1.5.0.4

Chapter 16

Protocols for Efficient Simulations of Long-Time Protein Dynamics Using Coarse-Grained CABS Model

Michal Jamroz, Andrzej Kolinski, and Sebastian Kmiecik

Abstract

Coarse-grained (CG) modeling is a well-acknowledged simulation approach for getting insight into long-time scale protein folding events at reasonable computational cost. Depending on the design of a CG model, the simulation protocols vary from highly case-specific—requiring user-defined assumptions about the folding scenario—to more sophisticated blind prediction methods for which only a protein sequence is required. Here we describe the framework protocol for the simulations of long-term dynamics of globular proteins, with the use of the CABS CG protein model and sequence data. The simulations can start from a random or a selected (e.g., native) structure. The described protocol has been validated using experimental data for protein folding model systems—the prediction results agreed well with the experimental results.

Key words Folding pathway, Folding mechanism, Protein dynamics, Protein folding, Coarse-grained modeling

1 Introduction

Protein folding events occur over a wide range of time scales: from picosecond (small fluctuations) to millisecond or longer (significant regrouping of thousands of atoms). No single experimental technique has yet presented a complete insight into the folding process due to the limitations in accessible time and resolution scales [1]. For small proteins, the 1 ms time scale has recently become accessible to atomic level molecular dynamics (MD) simulations run on special-purpose supercomputers [2]. Given the ambiguity of the experimental data, the major role of simulation techniques is to provide detailed structural models suitable for the experiment interpretation [3, 4].

The purpose of the described CABS software package is to perform long-time simulations of protein molecules including de novo folding from a random structure, near-native dynamics, unfolding processes, and long-time dynamics of unfolded structures.

Daisuke Kihara (ed.), *Protein Structure Prediction*, Methods in Molecular Biology, vol. 1137, DOI 10.1007/978-1-4939-0366-5_16,

The package simulation engine is the CABS protein model—a coarse-grained (CG) modeling tool—enabling an effective simulation of protein dynamics (at a much reduced computational cost compared to the most established simulation approach: an all-atom MD) and de novo prediction of protein structures. In the CASP experiments, the CABS-based prediction approach allowed for realistic de novo predictions of new folds for small proteins and an accurate modeling of large structures using various partial restraints derived from detected homologies with known structures (the approach was ranked first or second depending on the scoring system [5, 6]). The application of the CABS model to simulations of protein dynamics has been validated on experimental long-time scale (super-millisecond) data for protein folding model systems (perhaps the most extensively studied by experiment and theory): barnase [7], chymotrypsin inhibitor 2 [7], B1 domain of protein G [8], B domain of protein A [9, 10], and others [11]. The obtained simulation results concerning the folding mechanism or the denatured state properties agreed well with experimental data and other simulation findings (the review and comparison of the experimental, the CABS-predicted, and other simulation data for three protein folding model systems are presented in ref. 1). Another validation study included the comparison of the CABS dynamics with the results of MD simulations [12]. The test demonstrated that the consensus view of protein dynamics from short (10 ns)-time scale MD simulations (for different protein metafolds, using all-atom MD, explicit water, and four most popular force fields) is fairly consistent with the CABS dynamics. The CABS modeling approach has also been used in simulation studies of a chaperonin effect on folding mechanism (a simple chaperonin-like protocol was implemented within the CABS algorithm) [10].

Generally, in comparison with other simulation tools, the advantageous features of CABS include suitability for de novo prediction of small proteins, low computational cost of simulating significant conformational changes, and, in respect to other CG models, high resolution of coarse graining (physically realistic models can be obtained [9, 13]). The potential applications of the CABS model comprise structural characterizations of protein conformations along the folding pathway (denatured state ensembles, intermediates, and near-native ensembles) and thus the interpretation of the existing sparse experimental data. In all these prediction tasks, weak and/or fragmentary distance restraints (derived from sparse experimental data or from theoretical predictions of plausible structural biases) can be applied. Finally, the CABS-derived structures and trajectories can be used in multiscale modeling procedures, merging CG modeling with atomic level simulations (*see* the pipeline in Fig. 1).

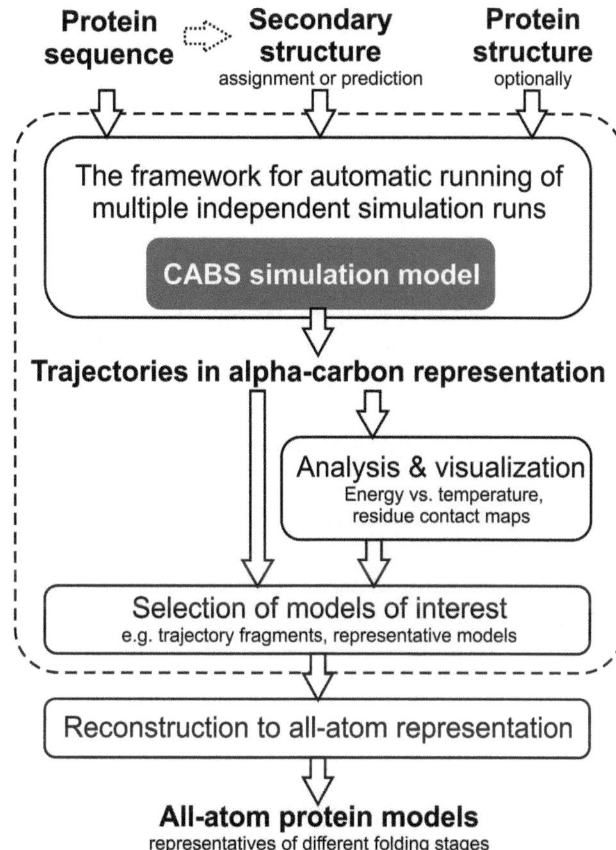

Fig. 1 Multiscale characterization of protein dynamics pipeline with the use of the CABS model. The framework protocol for simulation and analysis described in this manuscript is marked with a *dashed line*

2 Materials

2.1 Input Data

The required input data are protein sequence and assigned (or predicted) secondary structure. The optional input is starting structure data (in PDB format; required for unfolding studies or RMSD-to-native analysis).

For barnase, the example protein studied in the Methods section, the following data have been used: sequence, structure (PDBID: 1BNR), and secondary structure assignment (by the DSSP method) [14]. For known protein structures, both PDB and DSSP files can be accessed from the PDB database (http://www.rcsb.org/), e.g., the files for 1BNR can be obtained using the following links:

http://www.rcsb.org/pdb/files/1bnr.pdb.

http://www.rcsb.org/pdb/files/1bnr.dssp.

2.2 Software

The required software modules are the CABS modeling package and the CABS Python wrapper (pyCABS) both available for download from http://biocomp.chem.uw.edu.pl/pycabs. For running the pyCABS, Python interface and necessary Python modules (listed in Subheading 3) are needed. For additional download and setup details *see* **Note 1**.

2.3 Skills

The user should have basic skills in Python language scripting as well as a basic knowledge of structural bioinformatics (particularly in the foundations of protein folding problems and the use of protein structural data).

2.4 Hardware

A computer running Linux/Unix with at least 3 GB of free hard-disk space for the output data. Since some of the protocols described here involve running multiple (up to one hundred) simulation runs, we recommend the usage of a multi-CPU workstation.

3 Methods

The details of the CABS protein model are described in ref. [15]. Below, step-by-step instructions are presented together with python script fragments (given in `Courier New` font style). The complete scripts are available from http://biocomp.chem.uw.edu.pl/pycabs.

3.1 Environment Preparation

Download the required software (for download instructions *see* **Note 1**). Next, the necessary Python modules need to be imported. Create a file with the *.py extension (e.g., `folding_pathway.py`) and type inside

```
#!/usr/bin/env python
import matplotlib as mmp
mmp.use('Agg')
import os, random, pylab, glob, pycabs, numpy as np,
multiprocessing as mp
```

The first line is the information for the system which interpreter should be used for running the script. The next two lines define the environment for creation of contact maps and standard deviation plots. The last line invokes imports of the multiprocessing module (for parallel execution of CABS software), pylab (for plotting the data), and pyCABS (for running CABS and processing CABS format files).

3.2 Running Isothermal Simulations

The following example describes how to run multiple simulations of protein folding dynamics, for the example protein barnase. The described simulation approach was used in the characterization of the barnase folding pathway in the work of Kmiecik and Kolinski [7].

Note that the results may vary in quantitative details due to possibly different simulation settings and/or later modifications of the CABS model.

It is recommended, but not required, to provide sequence and secondary structure information using the DSSP file format (for additional hints *see* **Note 2**):

```
sequence,secstr = pycabs.parseDSSPOutput("1bnr.dssp")
```

Alternatively, one can simply define it in `sequence` (protein sequence) and `secstr` (protein secondary structure) variables. The secondary structure should be defined for each amino acid in the three-letter code: H, a helix; E, an extended state; and C, a coil (less regular structures). In the case of secondary structure predictions, overpredictions of the regular secondary structure (H or E) are more dangerous for the quality of the results than underpredictions.

In previous works, as the first step in the characterization of long-term dynamics we found it convenient to execute multiple isothermal simulation runs in different temperatures. In the CABS algorithm, the temperature is the parameter controlling the acceptance ratio for new conformations (through an asymmetric Monte Carlo scheme).

To run simulations in a parallel fashion (one simulation on one thread), create a function definition (`runCABS`) for the multiprocessing threadpool:

```
name = "barnase"
template = ["/where/is/my/barnase/1bnr.pdb"]
independent_runs = 5
temp_from = 1.5
temp_to = 3.8
temp_interval = 0.05
temperatures = np.arange(temp_from,temp_to,temp_interval)

def runCABS(temperature):
   global name, sequence,secstr,template,independent_runs
   here = os.getcwd()
   for i in range(independent_runs):
       temp = "%06.3f" %(temperature)
       dir_name= name+"_"+str(i)+"_T"+temp
       a = pycabs.CABS(sequence,secstr,template,dir_name)
       a.rng_seed = random.randint(1,10000)
       a.createLatticeReplicas(replicas=1)
       a.modeling(Ltemp=temperature,Htemp=temperature,
phot=300,cycles=100,dynamics=True)
       os.chdir(here)

pool = mp.Pool()
pool.map(runCABS,temperatures)
```

The above code fragment contains a declaration of the `independent_runs` variable, which tells the script to start five independent (with a different pseudo-random number generator seed) simulations for each temperature, starting from 1bnr.pdb (the native structure). It also contains the `temperatures` variable, which is a list of temperatures in the range of 1.5–3.8 and interval 0.05. This gives a total number of (3.8–1.5)/0.05 × 5 = 230 independent simulations (for additional details *see* **Note 3**). In order to start the simulations from extended random coil structures leave the `template` variable empty, i.e., `template=[]`. Doing so ensures that the simulation results are not biased from the starting structure. At elevated temperatures, due to the fast relaxation of the polypeptide chain, the simulation trajectory relatively quickly becomes independent from the starting structure.

The following parameters define the simulation length:

`cycles`—defines the number of CABS MC macrocycles [15] and determines the trajectory length (a number of trajectory snapshots is equal to `cycles` multiplied by 20, e.g., for `cycles=100` the resulting trajectory will have 2,000 snapshots).

`phot`—determines simulation length between the recorded snapshots.

The CABS-generated trajectories are produced in different output formats and representations: TRAF file (contains trajectory models in an alpha-carbon representation) and TRASG (contains trajectory models in a center-of-side-chain-mass representation). Both files are reformatted to a more popular PDB format. Additionally, each working directory contains an ENERGY file with CABS energy values for each model in a trajectory.

The CABS model (and the pyCABS module), developed primarily for protein structure prediction, enables application of distance restraints (derived from sparse experimental data or from theoretical predictions of plausible structural biases). For example instructions on running comparative modeling (with the use of structural template(s)), de novo modeling (template free), and modeling with the use of external distance constraints, *see* **Note 4**.

3.3 Calculating Simulation Statistics

Below are the instructions for the calculation of average CABS energy and standard deviation of energy values for the obtained trajectories. Both measures plotted in the function of temperature give an insight into the overall characteristics of the CABS energy landscape.

The standard deviation of energy (E) in function of temperature (T) is defined as

$$\sigma(T) = \sqrt{\frac{1}{N}\sum_{i=1}^{N}\left(E_i^T - \overline{E^T}\right)^2}$$

where N is the number of observables, and $\overline{E^T}$ is the mean in the given T.

To compute average energy and its standard deviation for each simulation, run the following code (`e_path` must be constructed identically to the `dir_name` variable in the `runCABS` procedure):

```
stdd = np.empty([independent_runs,len(temperatures)])
avgene = np.empty([independent_runs,len(temperatures)])
for j in range(independent_runs):
   for i in range(len(temperatures)):
     temp = "%06.3f" %(temperatures[i])
     e_path = os.path.join(name+'_'+str(j)+'_T'+temp,'ENERGY')
     energy = np.fromfile(e_path,sep='\n')[1000:]
     stdd[j][i] = np.std(energy)
     avgene[j][i] = np.mean(energy)
```

Adding the following commands

```
mean_sigma = np.mean(stdd,axis = 0)
stddev_sigma = np.std(stdd,axis = 0)
mean_ene = np.mean(avgene,axis = 0)
stddev_ene = np.std(avgene,axis = 0)
```

invokes computation of the average values from five independent simulations for each T value (*see* Fig. 2).

3.4 Plotting Simulation Statistics

To plot average CABS energy and standard deviation of energy values for the obtained trajectories (a single point denotes a single trajectory), users can apply the pylab module as in the code below

```
pylab.ylabel(r'Standard deviation of energy')
pylab.xlabel(r'Temperature, $T$')
pylab.xlim(temp_from,temp_to)
for i in range(independent_runs):
    pylab.plot(temperatures, stdd[i], '.')
pylab.errorbar(temperatures,mean_sigma,yerr=stddev_sigma,fmt='o-')
pylab.savefig("stdE_barnase.png",dpi=600)
pylab.close()
```

and analogously for the average energy plot (by changing `mean_sigma` to `mean_ene`, `stddev_sigma` to `stddev_ene` and `stdd` to `avgene`). The standard deviation of energy is written to `stdE_barnase.png` (upper panel in Fig. 2). The average energy plot is shown at the bottom of Fig. 2 (additional plotting options are given in **Note 5**).

3.5 Generating Contact Maps

Average contact maps (average for the entire isothermal trajectory or trajectory fragment of interest) provide a very informative insight into complex intramolecular interactions of highly diverse protein ensembles.

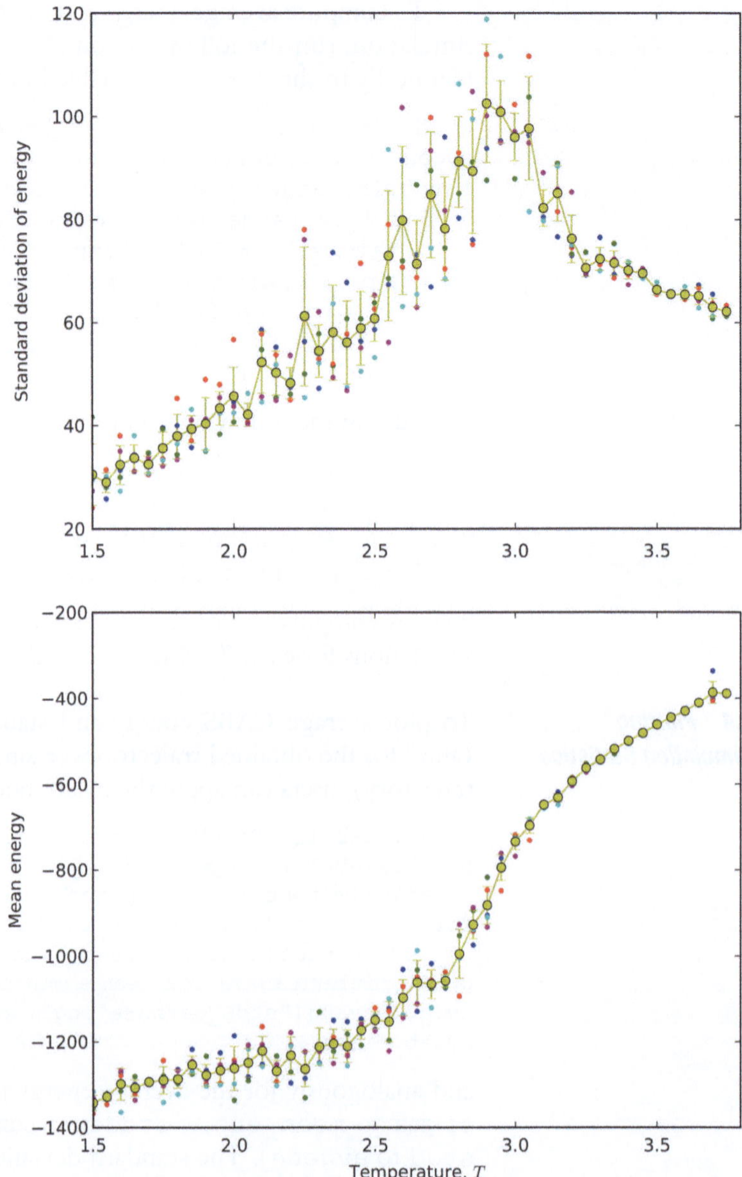

Fig. 2 CABS energy standard deviation (*above*) and CABS energy (*below*) as a function of temperature (*T*) for barnase (similar results were presented in ref. [7]). Variously colored small points represent individual isothermal simulations, while *larger yellow points* represent average value from five independent simulations in the given *T* value. The transition temperature (Tt) is identified by a steep drop of the energy and the peak of the energy standard deviation (heat capacity), here when *T*=2.9. Tt cannot be strictly identified with the transition state of protein folding. Sometimes, as for chymotrypsin inhibitor (*see* ref. [7]), conformations observed at Tt may be relatively unstructured, with some features of a molten globule state. For a more exact estimation of the Tt value one can repeat the computations in a smaller range of temperatures, with a smaller `temp_interval` value

```
#!/usr/bin/env python
import matplotlib as mmp
mmp.use('Agg')
import pycabs,os,numpy as np
name = "barnase"
max_sd_temperature=2.9
independent_runs=5
trajectory = []
for j in range(independent_runs):
     temp = "%06.3f" %(max_sd_temperature)
     e_path = os.path.join(name+'_'+str(j)+'_T'+temp,'TRASG')
     trajectory += pycabs.loadSGCoordinates(e_path)[1000:]
```

The above code fragment loads the second half of trajectories in the center-of-side-chain-mass trace format from five independent simulations in the temperature 2.9. To calculate an average contact map (the contact map definition is given in **Note 6**) with the cutoff of 7.0 Å use

```
contact = pycabs.contact_map(trajectory,7.0)
```

and to write it to a file, use the pylab module (for the map coloring hint *see* **Note 7**):

```
from pylab import xlabel,ylabel,pcolor,colorbar,savefig,
xlim,ylim,cm
xlabel("Residue index")
ylabel("Residue index")
xlim(0,len(contact))
ylim(0,len(contact))
for k in range(len(contact)-3):
    for l in range(3):
          contact[k+l][k+l]=contact[k+l][k] = contact[k]
          [k+l] = contact[k][k]=0

pcolor(contact, cmap=cm.gnuplot2_r,vmax=0.6)
cb = colorbar()
cb.set_label("Fraction of contacts")
savefig("average_heatmap"+str(max_sd_temperature)+".png")
```

The example contact map, created as described above, is presented in Fig. 3.

Note that for generating contact map figures, instead of using the pylab module, one can use any specialized software for this purpose, e.g., Gnuplot program (for plotting instructions in Gnuplot *see* **Note 8**).

3.6 Selection of Models of Interest Using RMSD-to-Native and CABS Energy Measures

The resulted trajectories can be filtered and structurally analyzed using simple filters (for example CABS energy and RMSD-to-native cutoffs). More sophisticated structural analysis is perhaps most commonly performed with the use of clustering analysis [16] (like in the characterization of near-native ensemble in ref. [8] or transition state ensemble in ref. [9]) or principal component analysis [17].

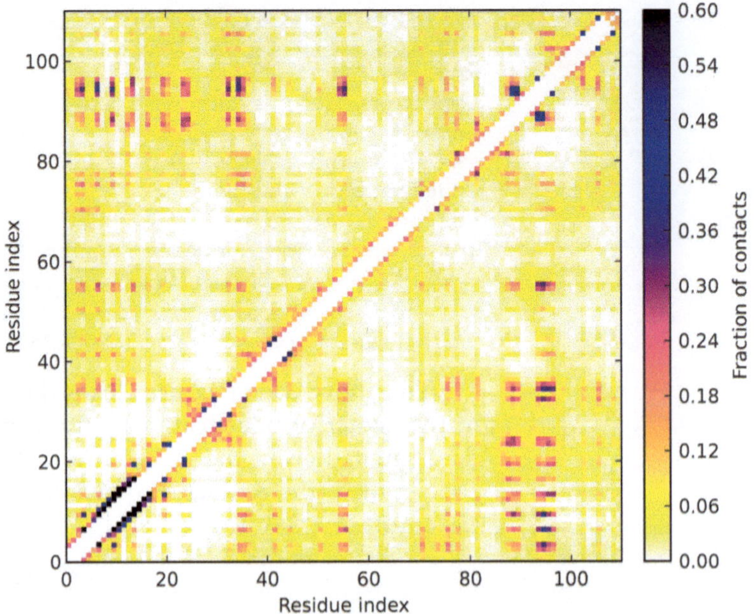

Fig. 3 Contact map for the intermediate (between fully denatured and near-native) state of barnase (similar results were presented in ref. [7]). The *colors* indicate the frequency of contacts. Short-range contacts are omitted for clarity

The simple filtering options are accessible from the provided modules. To filter out models dissimilar by RMSD to the native structure, one can use

```
#!/usr/bin/env python
import pycabs,os,numpy as np

name = "barnase"
rmsd_cutoff = 7.5
max_sd_temperature=2.9
independent_runs=5

native = pycabs.parsePDBfile("/path/to/barnase/1bnr.pdb")
trajectory = []
for j in range(independent_runs):
    temp = "%06.3f" %(max_sd_temperature)
    e_path = os.path.join(name+'_'+str(j)+'_T'+temp,'TRAF')
    for model in pycabs.loadTRAFCoordinates(e_path):
        if pycabs.rmsd(native,model) < rmsd_cutoff:
            trajectory += model
```

Note that each simulation directory contains an `ENERGY` file with the energy of each trajectory model. By reading it to memory (`numpy.fromfile("path/ENERGY",sep = "\n")`) the user can filter out models with a particular energy cutoff.

3.7 Reconstruction to All-Atom Representation

Selected individual models or consecutive trajectory fragments can be rebuilt to an all-atom representation. The task of reconstruction from alpha carbon trace is typically solved by a two-step procedure [9, 18]: backbone reconstruction from alpha carbon trace [19] followed by side-chain reconstruction [20] based on the position of backbone atoms. Note that models from the CABS method (in alpha carbon trace representation), as well as from the other CG modeling tools, are not free from unphysical local distortions. Therefore, building physically sound models from reduced models usually requires specialized reconstruction and refinement procedures [18].

4 Notes

1. All necessary applications can be downloaded from the following sources: Python (http://www.python.org), Pylab (http://www.scipy.org/PyLab), CABS/pycabs (http://biocomp.chem.uw.edu.pl), and GNUplot (http://www.gnuplot.info). All programs (except pyCABS and CABS) are available in most of the Linux distribution repositories.

   If one wants to compile the CABS software, use `g77 -O2 -static -ffloat-store -o cabs_dynamics CABS_dynamics.f` and move the `"cabs_dynamics"` file to the `FF` directory of pyCABS module.

   After downloading the pyCABS package, uncompress it into the working directory and modify the `pycabs.py` file by setting path to the FF directory. This can be done by changing the `self.FF` variable in the `__init__` method of the `CABS` class.

2. One can utilize secondary structure prediction software and write a predicted secondary structure (each residue in one-letter code: H—helix, E—extended, C—coil) in the `secstr` variable. Note that the Protein Data Bank does not contain DSSP files for all deposited proteins.

3. This task has taken about 28 h on 24 Intel® E5649 threads. That range of temperatures is typical for barnase; for other proteins it could be different. In order to roughly estimate the appropriate range, an initial simulation run can be performed with less computationally expensive operands: `temp_interval=1` and `independent runs=1`. Note that `pool = mp.Pool()` uses all available CPUs by default, but the user can limit it, e.g., `pool = mp.Pool(4)`, to utilize only four CPUs.

4. Example instructions for running: comparative modeling (with the use of structural template(s)), de novo modeling (template free), and modeling with the use of external distance constraints.

The following is an example script for protein structure prediction by comparative modeling (with the use of secondary structure prediction (in the Porter method [21] format file) and three templates (e.g., from Pcons Structure Prediction Meta Server: pcons.net): t1.pdb, t2.pdb, t3.pdb). Residues in the template structure files have to be numbered according to the target sequence alignment:

```
#!/usr/bin/env python
import pycabs

sequence,sec_str = pycabs.parsePorterOutput("/absolute/
path/to/porter.ss")
working_dir = "prediction" # name of project
templates  =  ["/abs/path/to/t1.pdb","/abs/path/to/
t2.pdb","/abs/path/to/t3.pdb"]
a = pycabs.CABS(sequence,sec_str,templates,working_dir)
a.generateConstraints()
a.createLatticeReplicas(replicas = 10) # create start
models from templates
a.modeling(Htemp = 2.0,Ltemp = 1.0,cycles = 20,phot = 100)
```

The script presented above: (1) parses the secondary structure prediction file (one can directly define sequence and secondary structure in `sequence` and `sec_str` variables, respectively); (2) creates distance constraints from templates; (3) creates 10 starting structures projected on the CABS lattice (iteratively from each template), which can be viewed in PDB file format in the `"prediction"` directory; and (4) runs CABS simulation with REMC and simulated annealing in the temperature range from 2.0 to 1.0 (typical values for comparative modeling).

In order to run de novo modeling (without the use of templates/constraints) one needs to (1) specify the `sequence` and `sec_str` variables, (2) leave the templates empty (i.e., `templates = []`) and comment out the `a.generateConstraints()` line, and (3) run CABS simulation with REMC and simulated annealing in the temperature range from 3.5 to 1.0, `cycles=100, phot=100,` and `replicas=30,` which are typical settings for de novo modeling. Note that de novo modeling is an extremely difficult modeling task and the difficulty increases with the protein length. Thus, the procedure may be suitable for small proteins preferably not longer than 120 residues.

In order to introduce some external distance constraints (derived from sparse experimental data or from theoretical predictions of plausible structural biases), one can manually add the distances data before running the modeling procedure:

```
misc = []
misc.append((1, 40, 15.4, 16.6, 0.5))
a.generateConstraints(exclude_residues=range(1,1000),
other_constraints = misc)
```

The code fragment presented above (1) excludes all automatically generated constraints (`exclude_residues` for residues 1–999) and (2) adds user-provided constraint between the alpha carbon atoms of the residues No. 1 and No. 40 with the constraint range between 15.4 and 16.6 Å (the constraint range is a preferred distance between the selected alpha carbons) and the constraint force equal to 0.5. The variable `misc` is in the format of a list of tuples (`residue_i_index, residue_j_index, lower_distance, upper_distance, force`). If one needs to change the global force constraint, it is possible to do so by providing a new value for `constraints_force` (default 1.0) in the `modeling` method, i.e.,

```
a.modeling(Htemp = 2.0,Ltemp = 1.0,cycles = 20,phot = 100,
constraints_force = 2.0)
```

If the script is successfully terminated, the prediction results can be found in the "`prediction`" directory (TRAF.pdb file).

5. At the script level, one can define output plot parameters, e.g., label sizes, colors, and resolution (more visualization examples can be found at http://matplotlib.org).
6. Contact map C is a $N \times N$ matrix defined as

$$C(i,j) = C(j,i) = \begin{cases} 1 & if\ d(x_i, x_j) < cutoff \\ 0 & otherwise \end{cases}$$

where x_i is the position of the x-th atom (here the center of a mass of a side group of an i-th residue).

7. The `Pcolor` function of the pylab module has a `vmax` parameter which defines the maximum value of the colorbar scale. Manipulating the `vmax` value may be helpful for a proper visualization of contacts of interest.
8. Instead of using the pylab module, one can write text data to the output file. To write the contact array into a file formatted for GNUplot, write a file with three columns (i-th residue, j-th residue, contact fraction value) and leave a blank row each time before the i-th column changes its value:

```
fw = open("contact_map.dat","w")
for i in range(len(contact)):
    for j in range(len(contact)):
       fw.write("%5d %5d %7.5f\n" %(i+1,j+1,contact[i]
[j]))
    fw.write("\n")
 fw.close()
```

Note that in the example above, the script writes residue indexes starting from 1 (in pylab fragment it creates plots starting from 0).

Finally, plot `contact_map.dat` in GNUplot (and write to the postscript file with a font size suitable for presentation):

```
set terminal unknown
plot 'contact_map.dat' using 1:2:3
set xrange[GPVAL_DATA_X_MIN:GPVAL_DATA_X_MAX]
set yrange[GPVAL_DATA_Y_MIN:GPVAL_DATA_Y_MAX]

set terminal postscript eps enhanced color "Helvetica" 14
set output 'contact_map.eps'
set size ratio 1
unset key
set xlabel 'Residue index'
set ylabel 'Residue index'
set cbrange[:0.8]
set palette negative

plot 'contact_map.dat' with image
```

The first four lines of these GNUplot commands are responsible for the calculation of max/min values of axis data (1 to chain length); `set cbrange[:0.8]` sets the colorbar scale in the range of 0.0–0.8.

5 Case Studies

Below are brief descriptions of several applications of the CABS model, together with the post-processing analysis applied to the characterization of protein folding.

A staggering number of different protein conformations sampled during de novo simulations require post-processing strategies that reduce the vast conformational complexity into easy to understand and interpret data. The complex nature of intramolecular interactions of highly diverse ensembles can be relatively simply described by average contact maps (average for the entire isothermal trajectory or trajectory fragment of interest). As shown in the folding mechanism studies, the characterization of the appropriate protein ensembles in the form of the averaged residue contact maps (derived from the trajectories in CG representation), matched very well with the experimental data from protein engineering (phi value analysis) [7–10]. The relative contact frequencies from the CABS simulations were also shown to be in semiquantitative agreement with experimental data (phi value analysis, hydrogen-exchange protection factors) [8, 10, 11] and other theoretical predictions [8, 12]. In the case of the B1 domain of protein G folding studies [8], quantitative analysis of the clusters of the most persistent native long-range side-chain contacts and their evolvement from highly denaturing to native conditions allowed for a detailed (residue–residue contact level) description of the folding events. Apart from the contact-level description of the highly diverse ensembles, some persistent conformers appearing along the folding route can be structurally characterized through clustering analysis (as shown for the ensembles of the transition state of the B domain of protein A [9], and the native-like globule of the B1 domain of protein G [8]).

Acknowledgments

The authors acknowledge support from a TEAM project (TEAM/2011-7/6) co-financed by the EU European Regional Development Fund operated within the Innovative Economy Operational Program and from Polish National Science Center (Grant No. NN301071140) and from Polish Ministry of Science and Higher Education (Grant No. IP2011 024371).

References

1. Kmiecik S, Jamroz M, Kolinski A (2011) Multiscale approach to protein folding dynamics. In: Kolinski A (ed) Multiscale approaches to protein modeling. Springer, New York, pp 281–294
2. Lindorff-Larsen K, Piana S, Dror RO, Shaw DE (2011) How fast-folding proteins fold. Science 334:517–520
3. Schaeffer RD, Fersht A, Daggett V (2008) Combining experiment and simulation in protein folding: closing the gap for small model systems. Curr Opin Struct Biol 18:4–9
4. Rizzuti B, Daggett V (2013) Using simulations to provide the framework for experimental protein folding studies. Arch Biochem Biophys 531(1–2):128–135
5. Kolinski A, Bujnicki JM (2005) Generalized protein structure prediction based on combination of fold-recognition with de novo folding and evaluation of models. Proteins 61(Suppl 7):84–90
6. Debe DA, Danzer JF, Goddard WA, Poleksic A (2006) STRUCTFAST: protein sequence remote homology detection and alignment using novel dynamic programming and profile-profile scoring. Proteins 64:960–967
7. Kmiecik S, Kolinski A (2007) Characterization of protein-folding pathways by reduced-space modeling. Proc Natl Acad Sci USA 104: 12330–12335
8. Kmiecik S, Kolinski A (2008) Folding pathway of the B1 domain of protein G explored by multiscale modeling. Biophys J 94:726–736
9. Kmiecik S, Gront D, Kouza M, Kolinski A (2012) From coarse-grained to atomic-level characterization of protein dynamics: transition state for the folding of B domain of protein A. J Phys Chem B 116:7026–7032
10. Kmiecik S, Kolinski A (2011) Simulation of chaperonin effect on protein folding: a shift from nucleation-condensation to framework mechanism. J Am Chem Soc 133:10283–10289
11. Kmiecik S, Kurcinski M, Rutkowska A, Gront D, Kolinski A (2006) Denatured proteins and early folding intermediates simulated in a reduced conformational space. Acta Biochim Pol 53:131–144
12. Jamroz M, Orozco M, Kolinski A, Kmiecik S (2013) Consistent view of protein fluctuations from all-atom molecular dynamics and coarse-grained dynamics with knowledge-based force-field. J Chem Theory Comput 9:119–125
13. Kmiecik S, Gront D, Kolinski A (2007) Towards the high-resolution protein structure prediction. Fast refinement of reduced models with all-atom force field. BMC Struct Biol 7:43
14. Kabsch W, Sander C (1983) Dictionary of protein secondary structure: pattern recognition of hydrogen-bonded and geometrical features. Biopolymers 22:2577–2637
15. Kolinski A (2004) Protein modeling and structure prediction with a reduced representation. Acta Biochim Pol 51:349–371

16. Jamroz M, Kolinski A (2013) ClusCo: clustering and comparison of protein models. BMC Bioinformatics 14:62
17. Maisuradze GG, Liwo A, Scheraga HA (2009) Principal component analysis for protein folding dynamics. J Mol Biol 385:312–329
18. Xu D, Zhang Y (2011) Improving the physical realism and structural accuracy of protein models by a two-step atomic-level energy minimization. Biophys J 101:2525–2534
19. Gront D, Kmiecik S, Kolinski A (2007) Backbone building from quadrilaterals: a fast and accurate algorithm for protein backbone reconstruction from alpha carbon coordinates. J Comput Chem 28: 1593–1597
20. Krivov GG, Shapovalov MV, Dunbrack RL Jr (2009) Improved prediction of protein side-chain conformations with SCWRL4. Proteins 77:778–795
21. Pollastri G, McLysaght A (2005) Porter: a new, accurate server for protein secondary structure prediction. Bioinformatics 21:1719–1720

Index

Daisuke Kihara (ed.), *Protein Structure Prediction*, Methods in Molecular Biology, vol. 1137,
DOI 10.1007/978-1-4939-0366-5, © Springer Science+Business Media New York 2014

H

I

K

L

M

N

P

MIX
Papier aus verantwortungsvollen Quellen
Paper from responsible sources
FSC® C105338

If you have any concerns about our products,
you can contact us on
ProductSafety@springernature.com

In case Publisher is established outside the EU,
the EU authorized representative is:
Springer Nature Customer Service Center GmbH
Europaplatz 3, 69115 Heidelberg, Germany

Printed by Libri Plureos GmbH
in Hamburg, Germany